WHERE IS THE WORLD GOING?

by Robert E. Barr

DORRANCE
PUBLISHING CO
EST. 1920
PITTSBURGH, PENNSYLVANIA 15238

The contents of this work, including, but not limited to, the accuracy of events, people, and places depicted; opinions expressed; permission to use previously published materials included; and any advice given or actions advocated are solely the responsibility of the author, who assumes all liability for said work and indemnifies the publisher against any claims stemming from publication of the work.

All Rights Reserved
Copyright © 2017 by Robert E. Barr

No part of this book may be reproduced or transmitted, downloaded, distributed, reverse engineered, or stored in or introduced into any information storage and retrieval system, in any form or by any means, including photocopying and recording, whether electronic or mechanical, now known or hereinafter invented without permission in writing from the publisher.

Dorrance Publishing Co
585 Alpha Drive
Pittsburgh, PA 15238
Visit our website at *www.dorrancebookstore.com*

ISBN: 978-1-4809-3796-3
eISBN: 978-1-4809-3773-4

CONTENTS

Preface		v
Acknowledgments		vii
Chapter 1	We Control Nothing	1
Chapter 2	Ancient History	3
Chapter 3	Human Beginnings	7
Chapter 4	Beginnings of Civilization	9
Chapter 5	Egyptian Rule	13
Chapter 6	Roman History Analogy	15
Chapter 7	Crusades	23
Chapter 8	RMS Titanic	25
Chapter 9	World War I	29
Chapter 10	The Great Depression	33
Chapter 11	Preamble to World War II	37
Chapter 12	World War II	41
Chapter 13	The Holocaust	49
Chapter 14	Auschwitz Concentration Camp	55
Chapter 15	World War II Japanese POW Concentration Camps	59
Chapter 16	World War II U-Boats	63
Chapter 17	World War II Japanese Submarines	67
Chapter 18	Messerschmitt ME-262	71
Chapter 19	Peenemunde	75
Chapter 20	Great Britain and Winston Churchill	81
Chapter 21	India and Great Britain	85

Chapter 22	Cold War	89
Chapter 23	Korean War	93
Chapter 24	Kennedy Assassination	97
Chapter 25	Vietnam	99
Chapter 26	Pueblo Incident	105
Chapter 27	EC-121 Downing Incident	109
Chapter 28	Iranian Hostage Crisis	113
Chapter 29	Beirut Bombing	115
Chapter 30	Airline Downing Incidents	117
Chapter 31	Gulf War	121
Chapter 32	World Warming	125
Chapter 33	Gun Control—Second Amendment	129
Chapter 34	How Large—the Government?	133
Chapter 35	DOD Budget	137
Chapter 36	National Debt	143
Chapter 37	US Oil Reserves	147
Chapter 38	World Government	151
Chapter 39	Socialism	155
Chapter 40	Fresh Water	161
Chapter 41	Learning from History	165
Chapter 42	Food Waste	171
Chapter 43	Common Core Education	175
Chapter 44	Celestial Havoc	181
Chapter 45	Apartheid	185
Chapter 46	Nuclear Weapons	189
Chapter 47	Religion and Science	193
	About the author	199

PREFACE

Civilization has arrived in the twenty-first century with technology and the advantages of the modern world. We have technical advances provided by science, but in terms of civilized judgment and philosophy, the last few thousand years contain evidence clearly demonstrating many errors in political, military, diplomatic, religious, environmental, financial strategy, and leadership. To commit a singular error in leadership is pardonable, but repeating these errors throughout years of possible learning is certainly not conducive to civilized progress. The world has had many conflicts from which we have learned little to nothing, as evidenced by two world wars and several world altering events. In the following chapters, there is an inclination to change our thinking from centering on leaders, special interests, wealth, power, or control, to preserving the world and humankind for progressive, enhanced, and contributing existences.

From history, some detailed examples are listed wherein significant leadership errors were made. Had other choices been made disregarding aggrandizement, power, control, wealth, selfishness, immorality, or godlessness, the world might have advanced civilization with free people at a better pace. Past world changing events contained philosophy and lessons of learning. Unfortunately, as humans, we have learned little of civilization and philosophy from the past. But there is always hope, whereby, earth's destiny can be brighter.

Acknowledgments

Before and during the initial stages of writing this book, my wife, Bettina provided inspiration, commentary, and critique. In the progress of the book, sadly, she passed away due to health issues that could not be overcome. She is greatly missed as a wife and mother.

With the experience and expertise of a World War II naval aviator, Roland Friederichsen provided reality and detail from a war which many people do not remember, a war that put world rule into question.

Neil and Shannon Fausey supplied commentary and critique. Shannon also provided the cover picture.

Don Butler provided an editorial review with commentary for enhancement.

WE CONTROL NOTHING

As we go through modern times with all the technological advances and problems caused by humanity, sometimes we wonder about human destiny. The reality is that humans literally cannot, even now, change the course of the earth in its 230 million-year orbit around our Milky Way galaxy. As intelligent and mighty as some leaders and segments of society believe they are, no mortal person nor group of people can change Earth's orbital track or destination. No matter how mighty or advanced our conception of ourselves might be, we are mere passengers on Earth. We might even be considered parasitic. Will we ever be truly able to control our destiny to include human peace, world unity, compatible governments, or governmental guidance for the world? From whence have we come? The journey to the present has been lengthy and complicated in each of humanity's epochs but much shorter in terms of universal history. As humans, we have been self-indulgent, only mindful of ourselves and not very cognizant of the health and destiny of the world on which we are merely passengers.

As we journey on Earth through time and space, we must realize that everything is relative. Scientists, geologists, and people of various disciplines agree that the earth is about 4.65 billion years old—very old compared to recent events of the last 5,000 years—but the earth, sun, and our local solar system are very young when compared to the universe's nearly 14 billion-year age. There are more celestial relationships then we can count. The earth and the sun are young relative to our galaxy, and we know very little of either. From the happenings of the last 5,000 years, we seem preordained to learning little of our greater universe. Xboxes and self-serving computers have advanced, but we remain self-indulgent—a human fault.

We have been observant of the heavens and celestial bodies for maybe 5,000 years and only in the last sixty years has mankind been proactive in exploring the cosmos. Lunar landings and satellites are evidence of greater human thinking. Affairs that are created by man such as politics, religion, rules, scripture, mandates, or financial functions, can all be adjusted, modified, or controlled by man. But geological happenings—catastrophes or environmental changes—and astronomical events are completely beyond human control.

Ancient History

Mankind's existence seems insignificant when compared to the initial formation of the earth 4.65 million years ago or the beginning of the universe 14 billion years ago. Earth came about 9.35 billion years after the initial formation of the universe as we know it. What is the significance of these numbers? Were they predetermined or happenstance? Many humans are not aware of cosmic ages, but the vast majority of humans believe in a supreme entity that guided creation. The probability of meaningful celestial formations, even during immeasurable spans of time, evades possibility without a guiding hand. These thoughts are not proof of divine guidance, but a functioning guidance— God looms imminent and necessary to the creation of a functioning universe.

Science and thinking man gave us the numbers with which we are working. Many religious zealots believe science and scientists are opposed to religion and the existence of a god, and some scientists believe religion is opposed to science. From a certain point of view these areas of endeavor may actually complement each other. Generally, fundamental religious practitioners do not believe that evolution brought the world, especially mankind, to our present state. Yet there seems to be a lack of appreciation for the eons of time utilized as a supreme being's tool in the evolutionary process. Time is limitless, although astronomers indicate time did not exist before the Big Bang and the universe's creation. Before any occurrences, however, something was proceeding.

From the time of the Big Bang there were 9.35 billion years before the formation of the earth, sun, and solar system, but our Milky Way Galaxy, in which we are residents, was formed at 13.6 billion years ago, shortly after or near the time of the Big Bang. This significant time span is about double the entire duration of Earth's existence. Of course, even as a thinking species, we can know nothing about vast eons of time; nor, can we really appreciate what occurred in that time. Galactic civilizations with durations of several billions of years could have come into existence—and disappeared. Before Earth's beginning, a civilization with a billion years or more of advancement and maturity could have flourished with technologies and lifestyles that we can't imagine. What we don't know is without measure, although we consider ourselves intelligent. We must realize and know that we don't know.

For considerably more than a billion years, the earth was in formation. External and internal conditions were at best catastrophic for lack of better descriptive words. Early in this formative period, probably 4.5 billion years ago, the earth and the planet Theia collided, which caused the earth to be larger, but enough ejecta was thrown into orbit to eventually coalesce into the moon. Two moons were possibly formed and existed for an undetermined time, and within an unspecified period, the moons collided to form Earth's current moon. During Earth's formation, the moon was much closer with a greater gravitational effect, but in 4.5 billion years, the moon has increased its distance to 239,000 miles from the earth, stabilized Earth's orbit, regulated the seasons, and the earth's oceanic tides. All of these functions have made life and human evolution possible.

As the earth traveled through the Precambrian supereon, the human mind had not evolved, and this span of 4.1 billion years cannot be imagined or properly described by any human being. This supereon contains the critical formation of life-giving oxygen on the earth. Current humans know little of the Precambrian and many are completely ignorant of it. This supereon was essential for the eras, epochs, and life forming steps that followed. Recently, geologists, paleontologists, and archaeologists have defined the time limits of the Precambrian, and the supereon represents eighty-eight percent of earth's existence—this is how little we know or control.

With the Precambrian supereon ending at 541 million years ago, the Paleozoic era began and lasted geologically until 252 million years ago. The Paleozoic era also ended the Permian period, which may have contained the greatest amount of biological die-off in all of Earth's geologic history. Whether humans choose to acknowledge the massive deaths during the Permian period or not is inconsequential in the present world. However, from 3.5 billion years to 252 million years in Earth's past, a lot of life forms had evolved, but the Permian-Triassic extinction eliminated all but five percent of all life during that time. The vast majority of humans are blissfully ignorant of their narrow escape from nonexistence. At the time, God's cosmic choice was to continue life with the remaining five percent. Humans would not be a factor or provide meaningful thought for another 250 million years.

Contained in the Mesozoic era are the Triassic, Jurassic, and Cretaceous periods. Evolutionarily speaking, our distant ancestors were present as proto-mammals, but their presence was insignificant compared to the dinosaurian animals that ruled the world for over 180 million years. In the three periods of dinosaur rule, no dinosaur appeared to possess intelligence beyond the daily living requirement activities if we judge by brain case. Until the end of the Cretaceous period, proto-mammals had progressed little in terms of intelligence and remained small and rather camouflaged from dinosaurs. From the beginning of the Triassic at 252 million years ago until the end of the Cretaceous period at 65.5 million years ago, paleontologically speaking, it would be difficult to say intelligent thought occurred at any time, but we don't know. However, at the end of the Cretaceous period, cosmic destiny did not bypass the earth. Scientific research by multiple, independent scientists and paleontologists have verified an asteroid, maybe the size of Mount Everest, collided with the earth near the Yucatán Peninsula in what is now the town of Chicxulub, Mexico. These areas are familiar to mankind now, since the continents were very near the current tectonic plate positions. Again, there were major extinctions but not of the magnitude of the Permian-Triassic extinction. This extinction was comprised of almost completely large animals, especially dinosaurs. Size was not conducive to life beyond the event. Large aquatic animals perished as did terrestrial beasts. Although it is certain Earth suffered, it gained from asteroids and comets prior to the end of the Cretaceous period. It is during K-T, the Cretaceous-Tertiary or Cretaceous-Paleogene boundary, when dinosaurs were removed as earth's inhabitants. Birds appear to be the only living remnants of dinosaurs.

From the geologic and paleontological evidence, scientists generally agree mammals could not have evolved to higher order speciation in the presence of the dominant dinosaurs. Although the Cretaceous period extinction was not of the magnitude of the Permian extinction, it was requisite for mammals, warm-blooded animals, and higher vertebrates to ascend to significantly higher intelligence levels. The scale of time does not seem to bear a great deal of importance, but the fact that the event occurred does bear the importance. It is difficult to say that a supreme being caused the Cretaceous extinction, but the event caused a meaningful and favorable direction for mammal evolution. For consideration, time and evolution could be tools of a supreme being. Looking from the present to the prehistoric past, humans have known so very little in the vastness of time in the limitless universe.

Human Beginnings

Paleontologists have been studying the fossils of early hominids for 200 or more years. The records have been inconclusive, paleontologists have thought definite answers were forthcoming, but additional fossils and evidence removed certainty. Current fossil evidence does not allow for straight lineage from present humans to Hominoidea, hominids, up to the last common ancestor of humans and apes. The road to modern humans has been complicated and filled with over twenty human species, including *Homo neanderthalensis*, a recent and near human competitor. Paleoanthropologists, a cross between paleontologists and anthropologists, have considered *Proconsul africanus* as a possible ancestor to greater apes and humans, but Proconsul existed 25 million years ago, near the end of the Paleogene period and during the Cenozoic era. However, our meager evidence demonstrates Proconsul lived prior to the ape-human split. At no time does evidence indicate humans evolved from apes or vice versa.

At what point did hominids or humans have control of anything? A clear answer has not been formed by paleoanthropologists. Control of anything must have occurred within the last few million years. If *Australopithecus robustus* of two million years ago is considered, there is some possibility of tool making, and in nearly the same era, *Homo habilis* used tools—a little control. At slightly less than one million years ago *Homo heidelbergensis* may have had crude tools or, possibly, spears for hunting. *Homo neanderthalensis*, 230,000 to 30,000 years ago, used crude tools, weapons, and appeared to be proficient at hunting. Although Neanderthals had some minimal control of hunting, they were slaves to their environment and the world. Neanderthals were not the direct ancestors of *Homo sapiens*, modern man, since Neanderthal DNA does not account for more than 2.4 to 4.0 percent of DNA in modern man. *Homo sapiens* appeared less than 200,000 years ago and coexisted with Neanderthals. At 40,000 years ago, Cro-Magnon man appeared, and is basically modern man. Cro-Magnon man used tools, cooked with fire, and protected himself with weapons. Yet, Cro-Magnon man controlled little of his environment, and he was reactive to fate and cosmic happenings. From Earth's and humans' early beginnings, civilization has only been upon us the last microsecond of known time, and even with civilization, humans still cannot control worldly happenings or the environment. Worldly control remains a term operational only within the realm of human imagination.

Beginnings of Civilization

The civilizations of Mesopotamia, in what is now Iraq, were initial attempts toward social order. In the valley between the Tigris and Euphrates rivers, man attempted civilization with huts, villages, and agricultural activities which may have begun in the valley as early as 12,000 years ago. More probable yet, civilized behaviors were established a few thousand years later. By gathering in the Mesopotamia Valley, humans had discovered that the power of collective, organized agricultural pursuits could feed society. The brain size of humans had grown to approximately 1,650 cubic centimeters, and the Mesopotamian fossils seem to portray long range planning, cognitive power. Eventually, more efficient agricultural means provided sufficient food for a class of nonfood producing people—aristocrats. Per worker, a surplus of food was produced.

There was a breakthrough in thought that concerned things other than survival and food production, all of this emanating from efficient agricultural methods. Services began to be rendered by a class of people not coupled to food in some way. Possibly, a hierarchy emerged where leadership was directing the actions of some humans, humans who had no thoughts connected to agrarian efficiency. The conditions were right for religion to be a concept or practice in man's activities. With a non-agrarian hierarchy, the individual freedom to conceive of religion and worship existed. The foundations of religion began. Since none of the monotheistic prophets had yet made an appearance, religion was centered about past ancestors, weather, world changing conditions, hunting, and the hopes for successful crop production. For those in the upper hierarchy of the rudimentary, small societies, religion of any method also produced a desirable societal byproduct of control, a psychological feature which has permeated society for several thousand years.

The monotheistic religions of Judaism, Islam, and Christianity had not yet produced the instigators of the religions: Judaism—Abraham, Isaac, Jacob, and Moses; Islam—Mohammed; Christianity—Jesus. These primitive religions were fundamental in forming the far more sophisticated religions that followed a few thousand years later, and in those years the prophets and key leaders did arrive. Judaism produced the initial leaders and foundations of belief, but no pure prophet was proclaimed. Christianity produced a prophet in Jesus Christ more than two thousand years ago. The teachings of Christ appear universally applicable to all humans and, if followed, religious involvement was a certainty. The teachings are universal throughout time, and as prophesied, the words of Jesus were given to him by God. Somewhat later, Mohammed was born around 570 CE in the holy city of Mecca, and at the age of forty, the angel Gabriel revealed Allah's word to Mohammed. Until his death, Mohammed was the self-proclaimed prophet of the Quran, the given word of Allah.

As the prophets emerged in world history and *Homo sapiens* began to become a more cognitive species—having gained dominion over agricultural production, lifestyles, and rules of religion—culture began to surface. Through religion and man-made compulsory rules, societal control began to be effective. From a commoner's point of view, the rules were in place to enforce compliance with God's will. Those in the thinking hierarchy were also enforcing rules which amplified their own societal powers. If the commoners did not comply, there would be God's, Yahweh's, or Allah's forbearance of future punishment or direct, vengeful, punishment as necessary. The sinners never realized that mere mortals had written the scripture for punishment. As time passed for the prophets or leaders of the monotheistic religions, tighter rules were imposed by the highest ranking religious leaders ensuring the maintenance of social order, maintenance of group discipline, unfailing obedience, adherence to religious dogma, and ostracism of those who questioned the rules of hierarchy. In the first few centuries after Christ, a few religions became all-encompassing to mankind. Intellectuals were forced to work in secrecy and fear of the death penalty administered by the church. A rise in the rank and authority within the church was not predicated on true knowledge of the world, physics, chemistry, or celestial relationships. With the development of the Roman Catholic Church, the Pope became the final authority on science, world geography, celestial discoveries, and the disbursement or teachings of various disciplines. The hard rulings on the religious dogma enforced by the Pope, eventually caused a break in the Catholic Church instigated by Martin Luther (1483-1546). With his movement of Protestantism, Luther was still not free from religious dogma, despite his learned background. Nearly any denomination of Protestant can demonstrate some good teachings established and written by mortals a couple of thousand years ago. Some of the teachings are beneficial to mankind, but many are aimed at control.

Before Martin Luther, there were as many as fifteen or sixteen crusades between 1095 and 1291, clearly indicating the desires for control and power. Many efforts were made to clear Jerusalem and the holy land of Muslims by Popes and the Holy Roman Catholic Church. Hierarchical powers along with the common man had consciousness, but they had not learned to think beyond their immediate world and environment. What power or influence could be exercised over conquered lands and peoples that was not tantamount to hierarchical thoughts? World saving environmental thinking had not occurred. Conquering instead of joining or mutual agreement was a way of advancement. Sadly, the last 150 years of civilization has not moved far from this line of thought. It is, indeed sad, as *Homo sapiens*, we are still bickering among our own members at this point in history.

EGYPTIAN RULE

What can be learned from the Egyptians (8000 BCE - 525 BCE), who established one of the oldest civilizations in mankind's history? The totality of Egyptian culture advanced science, art, religion, mathematics, and culture. Yet, through a myriad of rulers, Egypt fell from empire status to becoming a province of Rome (30 BCE - 476 CE), and, finally, was conquered by Arab Muslims in 646 CE. In a period of time better than eight thousand years, how does such a mighty civilization revert to directionless leadership and lack of motivation for national unity or sovereignty? Egyptian history was filled with rulers, too many to list, who built the Empire or those who were resting on the laurels of their predecessors. Succeeding Egyptian history did not always produce the vital leadership characteristics necessary for Egyptian growth. The Great Pyramid of Khufu has many dates assigned for starting and finishing. From dating methods in archaeology, geology, ancestry, flooding, or carbon dating construction dates are widely disparate. Comparing dates with ruler dynasties is not accurate either. The range of construction dates for the Great Pyramid spans from 3800 to 2285 BCE. Any further archaeological accuracy cannot be obtained.

If twenty years or more were required for construction, what outside factors affecting sovereignty and the good of the people were the Pharaohs considering? A first observation reveals any prosperity of the people was a byproduct of the king's singular quest for immortality through a lavish tomb effort. If the king fought wars for expansion or land, the efforts were to support agrarian needs of pyramid workers, who were probably not slaves. With the building of the Great Khufu Pyramid requiring at least twenty years of development, engineering, construction, geometry, astronomy, and transportation, great leadership skills were realized and brought to bear on the pyramid. As in previous and subsequent empires, the skills were focused on the singular, selfish goals of the king, not the advancement of the population or the country. While byproducts of prosperity were gained, they were not specifically designed for the greater improvement of the population or nation.

Evidence from the pyramids suggests that Egyptian disciplines were far more advanced than archaeologists, paleontologists, and Egyptologists suspected in the 1800's. Egyptian engineers and other builders certainly had equivalent intelligence to humans now. They did not have heavy machines or modern tools; yet they produced the pyramids, a construction which cannot be duplicated today. The movement and precise placement of cut stones weighing from two to fifty tons was well within their capabilities. Other singular stones or obelisks are far heavier, and the Egyptians managed producing and placing these behemoth stones wherever they desired. Unfortunately, Egyptian kings, leaders, and thinkers did not focus their talents toward promoting and prolonging the empire. For the self-aggrandizement of the king, the citizens and culture suffered. With more effort directed toward national growth and sovereignty, the Egyptian empire might have lasted much longer than it did. Yet realize that this observation comes from modern times where most nations have much shorter histories and national claims.

Are there comparisons to existing nations from ancient Egyptian kingdoms? Such comparisons are certainly subjective, but a few modern countries are providing far larger governments for control than is actually required for a prosperous, growing population. Some governments are the end product goal, rather than the catalyst for national and cultural advancement. Such observations are symptoms of politicians seeking power, rather than the results of regulating a culturally free, advancing citizenry.

ROMAN HISTORY ANALOGY

The vast and lengthy Roman Empire provides a broad view of history that lends itself to analogy to modern times and "civilized" societies. The empire lasted over a thousand years in an organized form. Stone tools, weapons, and pottery suggest a history of six thousand years for the people within what would be the future Roman borders. Despite many problems in Roman history, the empire stands favorable in longevity with modern countries. Rome became a Republic in 509 BCE, but the years of 300 BCE to 300 CE produced both power and corruption, as many rulers paraded through its lengthy history. By the last third of 100 CE, the city of Rome became one of the largest cities in the world with estimates of population between one and two million.

The country of Rome became an empire with power and influence to 300 CE, but problems were mounting internally and externally; as the empire grew rapidly it required deep expenditures for distant border maintenance. Heavy taxation fell upon Roman citizens and inflation became increasingly burdensome. Emperors and the elite spent gold for luxury items they perceived as necessities, depleting the country's gold supplies. Debt and inflation were financial factors strong enough to revert Rome to trade and barter.

Are these Roman mistakes lost in history or can analogies be drawn to prevent future civilizations from falling into the same pitfalls? Many factors were involved in the total decline of the Roman Empire, but a quick look reveals the analogy of the emperors' and senators' fiducial arrogance and their necessary responsibilities. From the republic to the formation of an empire, Emperors varied their financial responsibilities to include maintenance of tax monies to complete self-service of the emperor's wealth. Included in the Roman Senate were individuals involved in cronyism. These senators were totally involved with maintaining their senatorial positions and favor with the ruling emperor. The emperor prospered due to agriculture and trade with other nations, and for a lengthy period the empire grew economically despite the arrogance and self-serving attitude of the ruling elite.

Do we have countries for comparisons in modern times? The US is handy for inspection and could contain some faux pas that reveal a lack of learning from ancient or more recent histories. Past mistakes, ancient or more recent, cannot be rectified, but a learning process is warranted. The capitalistic and democratic system has proven itself to be the best governmental system in mankind's history, since the US has placed itself in history as the most powerful nation—ever. However, complacency, self-indulgence, and political aggrandizement can creep into high level thought regarding leadership. The nation tends to look the other way as elected politicians demonstrate interest in things other than the advancement and preservation of the nation. The forefathers did not intend for the Constitution to provide for professional, lifelong politicians or congressmen. A career in politics and increases in personal wealth were not intended consequences. As can be seen, Roman statesman followed the same behavior and pattern in reaping rewards from the government. Rome and the US have had statesmen who never undertook a full-time job in their respective workforce or country. It is difficult to lead or govern when one has never participated in the workforce, when only one's family forefathers actually worked for a living. In some cases, a family's ancestry made early fortunes allowing their offspring the luxury of eclectic lives, possibly in politics.

As the US grew from the colonies, a government was conceived that worked on checks, balances, credit, and debt. Notes were printed for credit and debt, which allowed the nation to grow based on a system honoring credit and debt. Trader or barter or the exchange of gold and silver ingots was cumbersome and slowed industry. A system of trust, honor, and reliability was necessary with a nation to back it. With ideals in place, a nation and its industry could grow rapidly.

Following the Romans, the US now has a class of statesmen who do not understand the financial system, either nationally or internationally. An easy rule follows: an individual, state, or country does not spend more money or markers of exchange than it has. Allowing debt to spiral to an irresponsible level beyond any near-term repayment schedule is ludicrous. For a nation, debt should be for expediency of commerce and trade, not indebtedness numbers to remain unattended and grow to unserviceable proportions. A nation that tolerates unserviceable debt is certain to encounter many problems—present and future. What kind of credibility is a country and future generations going to inherit? What kind of international trade can be expected from a country with insurmountable debt? The Romans had debt and inflation, which contributed greatly to the fall of the Roman Empire.

Presently, the US Congress operates imperviously to debt, or it seems so. Many congressmen act as if debt came from somewhere else and they are not responsible for it. But, Congress is responsible for the country's fiducial actions, and money management is one of those responsibilities. Paying debt is another responsibility. Yet, Congressman appear as if financial problems come from outer space and that God will solve their problems. The Romans thought likewise and that the future would somehow expunge their self-indulgent debt. Is the US government thinking similarly that the future will eradicate its debt and the responsibilities to it? Party affiliation makes no difference; the nation, as a whole, is responsible for its debt. Inflation and devaluation follows when a nation is not responsible or creditable for its behavior. For national longevity, the US and other debt ridden nations must rapidly address and began payments above the interest level of indebtedness. The Romans did not.

Roman wars were lengthy and expensive since Roman emperors spent great sums of money on constant warfare. To maintain and enhance Roman borders and to quell internal strife, armies were overstretched, requiring more soldiers. As a supplement, conquered barbarian soldiers were used, and, of course, loyalty to the Roman Empire was questionable. Raising, training, maintaining, feeding and otherwise expensing armies required expenditures to a point of producing unserviceable debt.

Does the US merit the comparison over the last 100 years? World War I and World War II purport exclusionary roles in the scenario, since the Allies were fighting for survival. However, a multitude of skirmishes not threatening the domain of the US have attracted the attention and intervention of US military forces—whether or not required. Some interventions may have been required, but all were not. Leadership is not a precise profession, and lack of good espionage intelligence can cause judgmental and leadership errors. Even with hindsight and the advantage of learning from recent history, it is questionable whether the US should have been in militarily jeopardizing locations. North Korea, Vietnam, and the middle east are battle areas in question. These conflicts have been costly in lives, materials, and expenditures. Statesmen argued endlessly and without resolution, leaving further international struggles as possibilities. To these conflicts and sovereignty disputes, there cannot be a singular answer satisfying all sides. For some conflicts, avoidance might have been a strong consideration. As with the Romans, the incurred debt might have indicated defeat even without a field victory. Heavy, long lasting debt cannot exist for continually successful nations—ever.

Certainly the Roman Empire reached greatness and power. During times when their fiduciary system was working and honored, ethics and moral standards were important in business and private affairs. The agrarian and trade systems had grown efficient, allowing more citizens to work non-laborious jobs. Some of these citizens rose to higher positions within the government, and a few saw opportunities to enhance personal wealth through compromises in ethics, morality, and values. Unfortunately, near the peak of Roman power, greed and declines in ethics were rife in business and personal affairs. Toward the end of the Roman Empire, 476 CE, the lack of ethics and morals had cheapened the life of citizens. Under such a continual decline of morals and ethics, no nation can count on national longevity.

Do modern times contain some similarities in ethics and moral decay to the Romans? Many countries in the world have ethics problems, but a look at the US might provide us with an example of ethics decline. Embarrassing questions must be asked to get an accurate profile of ethics here today. Why does the US have more prisons per capita than all other countries? Answers to this question have been couched in an immigration problem. Yes, some undesirables have crossed our borders, and this problem must be addressed. Most illegal immigrants coming to the US have self and family improvement in mind, and in time, they have been successful in fulfilling personal and family goals. From their backgrounds, food and clothing are major improvements compared to their past impoverishment. Without a doubt, undesirable immigrants pose moral and ethics problems, but these people are not fundamental to the national ethics problem.

Just as the Roman ethics declined with increased drug use, increased US per capita drug consumption has been inverse to its ethics for some time. For irrational reasons, some US states have legalized limited use of marijuana. Rational thinking cannot condone the open, legalized use of a mind altering substance—unless having a soft minded population is a goal.

With maturity as a nation, political corruption was operative within the Roman senate and through their weak emperors. Power and authority can grant favors to greedy, undeserving citizens and to organizations that might squander tax monies. Rome had lobbyists for senatorial influence, and for similar reasons, the profession remained in modern times and countries. Lobbyists are not necessarily bad, but their intent, generally, is to steer national monies toward private enterprise interest. The private concern may be well intentioned but probably does not have national sovereignty in mind.

Does the US have working, payroll entrepreneurial lobbyists? Yes, indeed, many corporations maintain high salaried lobbyists in Washington DC to influence Senators, Congressman, and political appointees. If lobbyists were not influential, to some degree, corporations and powerful individuals would have little need of them. With the use of lobbyists, where are morals and ethics placed, since the redirection of tax money is the ultimate goal? Records of the past would favor exchanges of ethics and morals for power, profit, and money. Since the establishment of civilizations and governments, lobbyists have worked for favor toward their supporting endeavors. In the process, ethics could be compromised.

Causes for the fall of the Roman Empire are numerous, nor are they simplistic. Christianity has been awarded some of the credit in Roman decline. For hundreds of years, Rome was a polytheistic nation, whereby many gods were accepted by the leaders and populace. The nation was uniform in its polytheism while Christianity interrupted the uniformity in worship. Additionally, Christians believed in a life after death. Before Constantine (324 CE - 337 CE), who proclaimed himself Christian, some Roman leaders may have feared Christians couldn't be punished because they believed in life after death. Maybe corporal punishment would not carry the fear as before Christianity. Initially, Christianity may have helped in dividing the nation, but believers in Christianity had no intention of dividing Rome. While Christians enjoyed their growth, followers may have had little insight to their political influence. By 380 CE, Theodosius established Christianity as the Roman official religion.

Emperor Constantine may have provided the largest factor in Rome's decline. During his reign that included 330 CE, he split the empire into two parts. The eastern part centered in Constantinople where Greek was spoken and Eastern Orthodox Christianity was the religion. The eastern part of the Empire survived as the Byzantine Empire for many years. The western part of the Empire spoke Latin and believed in the Roman Catholic Church. Economy failed for the western part of the Empire, although the city of Rome lasted for many years. Obviously, Constantine was not providing good leadership principles when he split the nation. The adage of "divide and conquer" was at work against the Romans. Other factors were causal for the decline of the Empire. Slave labor had a relationship to unemployment. The relationship is not direct, but a nation promoting slavery would affect free citizen employment. Hunger and disease may follow unemployment.

During Roman history, barbarian invasions were common, and at Rome's peak of power, barbarians were repelled. In the declining years with depreciating morals, ethics, and humanitarian values, Rome's corrupt leadership and army could not withstand the assaults. Some soldiers of the Roman army were former members of barbarian armies. Further, it follows from the standpoint of an internally weakened and sickened nation that solutions or corrective actions for the inevitable natural disasters were weak or nonexistent. At Rome's weakest point the nation could not respond to fires, earthquakes, flooding, etc.

If one were to look at the history of the US from the late 1700's, would the country's errors in leadership and sovereignty match or resemble those made in ancient Rome? In the comparatively short history of the US, the country has had a civil war over economics and slavery. Having slaves with civil strife belies ethics and morality problems. Of course, one human should not own another, but economics overcame ethics and morality. If enough people agree, then morality can be subjugated. The slavery issue is just one glaring comparison with ancient Rome. Several ethics and leadership issues of existing nations, including the US, have been mentioned as comparisons. These problems are not new and many are unresolved. Why? To remain sovereign, the US and other nations must communicate and begin solving problems critical to national longevity.

As mankind's history continues, our infinitesimally small time span is marked with demarcations of war, pitting humans against humans—supposedly rational, thinking creatures. If we continue like the Romans, religion will be more spiritual and much more of a contract with the forces of the world. Forces beyond Rome's control fell under the auspices of religion. Influences on the state and the military came from religion, and heads of households sought religion as a method of controlling both private and public affairs. A single deity or god was not recognized.

If Roman history is examined from the beginning of 750 BCE to the serious decline in 476 CE the Romans initiated many wars of acquisition and control through the avarice of the emperor's influence or, to some degree, through his personal religion. If Roman wars were motivated by expansion and control, many emperors' wars were fought with vigor. In the waning years of the Roman Empire, German barbarians, Franks, and Goths were formidable enemies of an empire declining from within. Leadership, especially the upper class citizens, had lost sight of national goals in life factors for the Empire. Initially religion had established rules and morality for the masses, but morality had decayed below a civilized measure. The unwarranted accumulation of wealth, passions for luxury, laziness, lack of motivation, slave labor, and declining patriotism were all strong enough factors to end nationalism. The end lineage of emperors exhibited no foresight, leadership, or strategic planning, but they were self-indulgent and self-serving, providing a clear path to national collapse and leadership failure.

To exemplify the past Roman Empires as major consumers of the world's environment and life-sustaining resources such as rivers, forests, soil, minerals, and the atmosphere would be astute. No Emperor thought of directing control of the resources toward conservation or preservation of future resources. For the time, a good Emperor thought of expansion, control, power, and basic requirements for the increasing population. Humans had developed brains and minds but they thought in an introverted style, which cared little about the destiny of the world. All things were natural while vigorous consumption was a natural duty of a nation. During the middle Roman era, politics became quite advanced as men of written, oral, and legal disciplines became powerful and influential before the Roman Senate. Educated and experienced Roman senators could and often did become self-serving, aggrandizing, and secular servants of power. Certainly, senatorial thought was not always for the national good and leadership— maybe it was good for corruption. The highest thinkers in the world still could not see man's relationship to the earth. Humans were in a race for consumption of power and empiric expansion. Man was only concerned about man. Long-term, earth saving promulgation was many years in the future, if ever.

CRUSADES

History records nine Christian crusades emanating from the Holy Roman Catholic Church, but more are probable. Why did Pope Urban II and high ranking Christian officials find the need to form a largely untrained army to conquer the holy land and the city of Nicaea, Dorylaeum, Antioch, and Jerusalem? A papal rethinking, political unrest, and the confrontation between Christianity and Islam involving many people were situations requiring solutions—more Christianity was needed according to the Pope. With any success the Roman Catholic Church would exercise control over more people and area. Through the church and the Crusades, the Pope wanted to quell violence in the east while assisting the Byzantine Empire. The rise of Islam, especially in Jerusalem, was not good propaganda for the papacy. An eradication of the Seljugs, Orthodox Sunni Muslims, in Palestine might further help in reuniting the churches of Rome and Constantinople.

Records are not accurate, but the sizes of the Crusaders' armies range from 35,000 to 60,000 eager participants. For the first Crusade in 1096 CE the logistical support was gargantuan; it required follow-up and local support to supply the needs of the advancing army. Subsequent waves of the Crusades occurred between 1098 CE and 1101 CE for helping the Byzantine Greeks in Constantinople. The catalyst to the first crusade was the request of Byzantine Emperor Alexios Komnenos to Pope Urban II asking for crusades to fight against the Seljug Turks in Asia Minor and Normans occupying the Balkans. For the request, Alexios sent envoys to the Council of Piacenza. Pope Urban

II cooperated with Alexios by dispatching far more first crusaders than Alexios expected. The Byzantine Empire was not logistically ready for the essentially untrained, poor masses that arrived. To the disadvantage of Alexios, the crusaders began pillaging Balkan villages. Alexios solved this problem by having their leader, Peter the Hermit, redirect the crusaders to Asia Minor and a temporary defeat by the Turks.

Were the objectives of the Crusades accomplished? In the short term the goals were attained. Were goals attained worth the lives lost, broken families, missing persons, pillage losses, and expenditures? A question of worth arises. For the Roman Catholic Church and Pope Urban II, Christianity was reestablished—if only temporarily. Depending upon who asks the question of were the Crusades worth it, one must measure the answer in terms of lives, morality and the reestablishment of Christianity. In retrospect, the years following have seen the growth of Islam in the areas covered by the Crusades, adding questions about foresight, leadership, and wisdom to the proponents of the Crusades.

RMS Titanic

Many accounts, books, and movies have been made about the sinking of the "RMS Titanic." Compared to ancient happenings, the sinking took place in relatively modern times. However, the air of invulnerability still mocked powerful nations when caution, rational thought, and common sense needed to prevail. While Captain Edward Smith was the most experienced captain of the White Star Line, he exhibited an attitude of overconfidence and lack of insight to the obvious and to nature. With little doubt, Captain Smith was time-constrained with a sea trial, certification, readying a crew, outfitting, and getting passengers aboard. Many of his high-ranking crew members were transferred from other ships, such as the "Olympic," and were changing duties. Only a very few had sailed aboard the "Titanic" and only on the sea trial. Crew confusion was bound to ensue. Captain Smith was pressured by his own pride and the White Star Line's quest for success on the first revenue launching of its much publicized luxury liner, "Titanic." Captain Smith was aware of the obstacles for the cruise: the first oceanic crossing for the "Titanic," making speed for timeliness, winter icebergs, only sufficient lifeboats as required by law but not per the passenger list, and an uncoordinated crew.

While Captain Smith had his worries, so did the White Star Line and its chairman, J Bruce Ismay. Ships are not impressive when docked for long periods, and without doubt, Mr. Ismay and the White Star Line needed to be timely in all operations. Protocol demanded that Ismay not influence Captain Smith in decision making, but the Captain was certainly aware of his presence.

It is not known, but some directives between Captain Smith and Ismay may have taken place. Certainly the two individuals conversed. Ismay mandated success for the White Star Line as did JP Morgan, who controlled International Mercantile Marine, IMM, White Star Line's parent corporation. Ismay was also president of the International Navigation Company, a constituent of International Mercantile Marine. The pressures to succeed from two countries, the US and Great Britain, were on the White Star Line, Ismay, Captain Smith, and the "Titanic."

JP Morgan as well as Jay Bruce Ismay badly needed the "Titanic," the White Star Line, and IMM to operate in an outstanding manner. If the "Titanic" operated otherwise, IMM and those in charge would be situated for great losses of money and positions, especially Ismay. Captain Smith understood these factors, and it is difficult to believe he did not operate the "Titanic" under these tremendous influences. In 1914 charts were available for the appearance or absence of celestial bodies, especially the moon, and as with other captains of North Atlantic vessels, Captain Smith knew of icebergs and field ice but yet maintained an attitude of invulnerability. Ice could not bother such mighty vessels as the "Titanic" or the "Olympic"—they were unsinkable. An old salt such as Captain Smith should have known that anything made by man can be overcome by nature, even if Harland and Wolff with Great Britain's blessing thought otherwise.

Although influenced by his own experience and Great Britain's Board of Trade, Captain Smith and others lacked common sense when comparing liner passenger capacity to lifeboat capacity. The prevalent idea was that lifeboats were to carry passengers from a foundering vessel to a nearby rescue vessel. But a poignant factor was overlooked—what if there was no timely rescue vessel? Great Britain had many ships—one must be nearby. This error of passenger carrying capabilities in the vastness of the sea had been carried for some time. One cannot help but think profit had overcome the expense of safety in the deficient lifeboat policy.

In fact, two nations were aligned with the deficient lifeboat policy. Britain's Board of Trade proclaimed ships over 10,000 tons needed only sixteen lifeboats, regardless of passenger count. The "Titanic" carried twenty lifeboats, more than the law required. And since the White Star Line used the Port of New York and carried US passengers, US maritime laws were aligned with those of Great Britain's. Each country wanted profitability. Captain Smith's providence was to sail—on time. He and other captains had used cruising speed through ice fields for some years, and such a procedure was standard. On the night of the April 14th, other factors became more prevalent: there was no moonlight to shine on the ice, calm seas prevented foam at the bottom of icebergs, continued cruise speed, no deviation—rhumb line navigation, and no sea—iceberg contrast for lookouts. Regardless of the decision, the radio shack prioritized passenger traffic over wireless ice field position warnings. The "California" had transmitted a warning and was shut out by the "Titanic" radio operators. Captain Smith, under the adverse circumstances, should have prioritized the radio shack for safety messages, but rich, self-important, overbearing passengers were allowed precedence to monopolize radio traffic—a dangerous, fatal procedure.

With the intercontinental passenger liner service becoming an advantageous mode of transportation and the promise of profits to the participating countries, identical, realistic safety regulations should have provided common safety pursuits. But, US Trade Board regulations for oceanic vessels were in line with the British Board of Trade regulation—unfortunately. Profits should not have gotten in the way of passenger oriented safety rules for the Secretary of Commerce or Transportation—for any country. If the "Titanic" had its full capacity of 3,547 passengers, and the lifeboats were utilized to the capacity of 1,178 its passengers, then, a possible death toll of 2,369 was possible with the rules. Fortunately, the "Titanic" was well below capacity due to a coal strike scare in England. The singular reason for its sinking was that it hit an iceberg. A myriad of circumstances could have altered the catastrophic historical result including: delayed sailing, slower speed, stopping in an ice field, adherence to wireless ice warnings, a better trained crew, and common sense on lifeboat capacity. The world had lost sight of safety in lieu of expediency and profit.

World War I

In human history, many wars and battles have occurred due to lack of leadership, insight, and the quest for individual or national advancement. World War I exemplifies a grand-scale lack of rational thinking on the part of its leaders. Mankind created a war lasting from July 1914 to November 1918, which had after effects until the depression. Lack of reason and intellect caused over nine million combatants and seven million civilian deaths. Two great combinations of forces were drawn into the conflict: The Allies consisting of the United Kingdom, France, with the Russian Empire and the opposing fascists' central powers of Germany and the Austro-Hungarian alliance. With this combination, over seventy million military combatants participated.

A catalyst to start the war can be found in the assassination of Archduke Franz Ferdinand of Austria, an heir to the throne of Austria-Hungary. Gavrilo Principe of Sarajevo was the culprit. When the Austria-Hungary alliance gave the Serbian kingdom an ultimatum, a diplomatic crisis ensued. At the time, international communication and alliances functioned in estranged and entangled manners, but with the incident, stronger alliances were quickly formed. Lack of communication coupled with erroneous beliefs brought nations to war. Certainly, some leaders lacked depth of reason for a world war. The prominent driving force centered on individual power, rather than notions of kingdoms standing for freedoms, democracy, and individual rights. Leaders, through blood lines, inherited reason and international judgment rights beyond their mortal capabilities. These inherited capabilities dominated in Europe. The

Germans, for example, violated The Hague Convention by using its technologically advanced chlorine gas. Was this violation for the good of the nation or to a further treacherous action to keep the leaders of the central powers in their position? As the war encompassed the world, the technology of tanks, machine guns, airplanes, battleships, and submarines advanced in efficiency for the maintenance of a ruling class.

Although not possessing the mightiest fleet of submarines, German leadership was willing to spend large sums of money to manufacture and maintain a submarine fleet. The U-boats were designed to cut off supplies from the US to Great Britain with early attacks coming without warning. High seas mayhem resulted for merchant shipping. The US protested the sinking of commercial ships, and after the protest, for a short time, U-boat crews allowed the crew and passengers to board life boats before the sinking. The sinking of the "RMS Lusitania" stirred international consternation, causing the Germans to promise not to sink passenger ships—temporarily. But commercial shipping began using convoys that enjoyed success in supplying Great Britain; it became clear to the Germans that the US would soon join the Allies. With these facts and US involvement, the German submarine cadre stopped restricting which ship they targeted. There were tragedies at sea for the ruling leaders and bloodline. In the World War I situation, the central power's self-indulgent thinking cost many lives. Possibly, from earlier years of Roman elitist thinking, German leaders considered themselves above the agrarian society and were exclusive to such thought.

As a necessary tool of the British and counter to the central powers of Germany, Austria-Hungary, the Ottoman Empire and Bulgaria, the "HMS Baralong" was converted from a cargo liner to a "Q" or special service vessel. The "Baralong" was transformed in 1915 by equipping it with three twelve-pound guns. To carry out its role, the ship was to be a wolf in sheep's clothing. Armament was camouflaged, and the deck had devices for simulating damage.

On August 19, 1915, the "Baralong" was responding to a distress call of the liner "Arabic," a 15,800-ton White Star liner that had been attacked and sunk by U-24. After several hours of pursuit, the "Baralong" encountered the British steamer "Nicosian," which was under fire from U-27. With trickery, diversion, and flag switching, Captain Godfrey Herbert sank the U-27 as it emerged from behind the bow of the "Nicosian." Aboard the "Baralong" there was a strong air of revenge due to the ruthless U-boat sinking of the "Arabic" and the "Lusitania." As U-27 sank and German sailors tried to swim toward the "Nicosian," crew members of the "Baralong" machine gunned the sailors in the water. Sailors that actually made it to the "Nicosian" were killed.

In a second incident with the "Baralong," now commanded by Lieutenant Commander A. Wilmont Smith, the ship performed a sheep-wolf act on U-41, which was attacking the "Urbino." As the "Baralong" approached U-41 after it had resurfaced, the "Baralong's" deck guns opened fire sinking the submarine and commander. Two German sailors escaped the sinking and were taken prisoner. One prisoner, first officer I. Compton, was returned to Germany. Of course, Compton made serious allegations against the British Navy that included treaty breaking actions and inhumanity on the high seas. The British Admiralty refuted all charges against the "Baralong," but the stories, however manufactured, made for reading in US and central powers papers.

Considering these instances and others of land, sea, and air in World War I, one wonders how such times managed to fester such catastrophes except by means of human ignorance. Certain periods of human history belie human intelligence. To counter ignorance, otherwise intelligent nations and leaders must acquiesce to methods equal or superior to a conquering power. Of course, ignorance for a conqueror does stand alone as a factor. Motivated self- aggrandizement and national aggrandizement must also be present.

World War I certainly contributed to the concept of ethnic cleansing or the idea of a pure race. Considering human history, World War I resides in fairly modern times; yet, leaders of the era had not developed a "human" race concept. Due to ignorance in leadership and patronage to emotions anchored to unsubstantiated societal rumors and hearsay, many people suffered unwarranted deportation, persecution, or death. World War I was not the first major conflict to patronize ethnic cleansing in the history of humanity. However, the Ottoman Empire paid undue attention to human factors and superficial differences. Intellectual value did not present a factor.

For a link to World War II postwar, the German discontent and rejection of defeat rose as a nationalistic attitude. Psychologically, Germans cast off the responsibility of defeat, and the general populace saw themselves as victims. A non-distinct, verbose, corporal named Adolf Hitler took advantage of this nationalism to sway masses towards fascism and a future utopia for Germany. Gullibility of the masses describes the situation. Well into the years following World War I, perceived injustices of the Allies justified

German acts of aggression. Hitler fully utilized this animosity against the Allies and victors for years into the future.

After the war, what did governments of the Allies and central powers do? For the Allies, politicians wanted more governmental power and responsibilities. New administrative offices were created. Of course, more people would be required, and a more than commensurate tax would be levied for support. Logic and logistics tend to support less government, not more. The administrative load added to the tax burden and was not conducive to a smoothly functioning federal government but was conducive to its further accrual of national debt, which a sovereign country should not tolerate.

The gross domestic product for the central powers decreased. Debt for Germany was burdensome, and through the Treaty of Versailles, an imposed payment schedule from Germany and its allies began. With the rise of Hitler, the repayments were forgotten for a lengthy period. By the middle of the 1930's, Great Britain owed the US in excess of four billion dollars, which was never repaid.

THE GREAT DEPRESSION

A good look at the monetary system of the US and its trading partners is warranted. As businesses progress, there are increases and decreases in market inflation and deflation. Many factors can influence the stock market, which are unexpected by gurus of trade and finance. Sometimes learned investors behave as if they were gerbils who asked questions of others after jumping over the financial cliff. The 1929 depression has as many explanations in terms of dollars of capital lost. With the Federal Reserve not slowing or stopping deflation, farming in a drought, union problems, rampant unemployment, and unstable gold reserves, a recession would clearly follow.

Primarily, lack of confidence by the working people and investors provided motive force of withdrawal from banks to begin failures. Furthermore, the Federal Reserve did not back the banks. If an institution is providing the financial backing of banks, it becomes difficult to win the war when community banks are not backed or supported. Lacking confidence, the common citizen is going to the bank for withdrawals. Shortsighted federal procedures were capping monies the Federal Reserve could have extended to banks for support while lack of confidence withdrawals occurred.

Citizens and investors reasoned: if the government can't control a downturn in the economy, which occurs often, then banks are not a good place for money, and cash is king. The mattress becomes the safe haven. As the idea amplified, industry, employment, and the economy suffered. If confidence in the economy is eroding, why put money into a losing microeconomic US structure? Certainly, the citizen saver of the 1920's and 1930's did not work with financial formulas or the internal rate of return on companies, and investors were interested in a price increase on their acquired stock. Or investors might buy a lower price stock if they were confident of future price increases and dividends.

While lack of confidence precipitated consumer bank withdrawals, the Federal Reserve allowed the Bank of New York to fail, moving financial thoughts into a panic. Banks were fine for mortgages and loans until the lack of confidence in the financial system precipitated massive withdrawals for safety. The US financial machine was similar to a nuclear meltdown. Investors and savers did not have the reason or confidence to invest and build. The Federal Reserve, by law, was required to back notes issued by forty percent gold reserve, and the Federal Reserve had neared its limit on notes issued covered by the gold in reserve. During the depression, the Federal Reserve redeemed some notes for gold, but with the onset of financial panic, the effort was late and well short of monies for failing banks.

International trade was affected by the depression in the US, and soon many countries lost confidence in the US banking system and eventually their own on the macroeconomic scale. In short term thinking, these countries, including the US, went for protectionist schemes such as tariffs. As each country capitulated to tariffs, the schemes failed and the macroeconomic situation worsened.

During the days of the depression, Congress and financial gurus acted as if our financial problems and woes were naturally caused. What had the country done to deserve such punishment? From history we know that mankind invented financial systems whereby trade, international or domestic, could be conducted based on confidence of credit or debt markers. With a system in place, progress and advancement in civilization was possible. So, mankind invented systems of monetary trade where credit and debt are recognized in a fiduciary system. Confidence and honor are two basic components of the system. Furthermore, it is recognized that more than a single system may exist, yet the system can mix and exchange equivalent values.

In the US, Congress makes the fiduciary law and appropriates monies as required. The Federal Reserve and the banking conglomerate practice distribution, credits, and debts. While the Federal Reserve or central banking was created by Congress, it is independent. However, the Federal Reserve is still accountable to Congress.

Other countries have similar banking systems, and as countries become allies, mutual honorariums are created in the international fiduciary system. Under this system and as credit or debt problems arise, somewhere in the complex of the financial machine, human intelligence has erred. We are responsible and there is no reason for Congress to look outside of itself for causal events. Theoretically the Federal Reserve should provide an accounting to Congress.

Preamble to World War II

After the world had endured World War I and the depression beginning in 1929, Germany, Japan, Italy and the US at the very least, were faced with retaining sovereignty of their nations. Through these times until 1939, Hitler and German leaders noticed that the world's biggest Empire was Great Britain. For some years, the British Empire had been growing, and not to the liking of its many subject territories. The British Isles of the UK had a stronghold on the world, technically not a direct problem but definitely a future sovereignty problem for aspiring dictators. Hitler could not help but notice the British expanse, whereupon the sun never set. A repercussion of Germany losing World War I was a shortening of its borders, which was insulting to all Germans, especially, to leaders who had expansion and empire on their minds.

From Churchill's point of view, the Soviet Union and communism were getting to be unnecessarily prominent and foreboding. To Great Britain's benefit, a Hitler-conquered Soviet Union would be beneficial, since Britain was already the world's largest empire. As far as Churchill was concerned, a crushing of the Bolsheviks was a step forward for civilization. Churchill was quite aware of the Jewish problems within Germany and surrounding countries, but he saw the Jews as a race, not a culture of people following a religion. As such, Churchill was not against anti-Semitism.

The closing of World War I saved the world from further human slaughter—temporarily. Ironically, Japan was allied with Great Britain during World War I. But, countries change their allegiance for economic reasons, and Japan was no exception. The 1929 financial collapse was a disaster for Japan, since the small island empire depended upon outside sources for materials and energy, especially oil. Japan had no oil internally and the financial collapse made outside acquisition nearly impossible. Under such circumstances, how long can a nation endure? Since the late 1800's the US had an "open door" policy with China, and Japan was aware of this trade agreement; the Japanese had the motivation to attack Manchuria in 1931 and Beijing in 1937—hostile, non-cooperative, anti-international hostilities.

Through the 1930's, actions and motivations of the fascist countries, Germany, Japan, and Italy, began to form a picture of an economic war for future sovereignty. US leaders and the intelligence people were aware of the fascists' warring potential. The questions for the Allies and the US intelligence specialists were who, when, where, and how. The question of when was partially answered as the US placed an oil embargo on Japan in July of 1941. Clearly, US intelligence estimated a curtailment of the Japanese oil supplies to within months, maybe less. The Allies were aware of the fascists' alliance, and Hitler had already attacked Poland. With the Japanese embargo in place, it was a certainty the Japanese would do something in the Pacific area— and soon. Within the Pacific arena, the only country that could stop the Japanese from raiding along the Pacific Rim was the US, since Great Britain was already at war with Hitler. The when part of the question was narrowing, and the US was a certainty for the who part of the question. Of course, a post- hoc analysis has advantages, but in the entire Pacific arena, Pearl Harbor contained a large naval fleet, capable of stopping Japanese aggression anywhere. One need not ascend high up the intelligence totem pole to determine the imminence of an attack on the Hawaiian naval fleet. Military intelligence and tacticians failed to forecast an air attack by Japan, and for that error, we paid the full price. However, the attack brought solidarity to the nation—the hard way.

Prior to Japanese aggression, operation ORANGE was instigated as an offensive and lengthy plan to impose US will on imperialist actions of Japan. ORANGE was the code word for Japan. Essentially, the plan would destroy Japanese armed forces and their capabilities to wage war. Implemented early enough and administered correctly, the Japanese attack could have been averted, but this conclusion was not warranted.

After the war had begun, the homeland of Japan evolved many Imperial fanatics, who would not surrender under any conditions. US military leaders pondered the Japanese fanaticism but, unfortunately, drastic measures were part of the solution in early 1945. The infrastructure of Tokyo was known to be weak, since material resources had been spent on war efforts. With the high percentage of wood structures, a firebombing of Tokyo would be most devastating, and post war facts revealed catastrophic damage to structures and personnel. In just a few hours after the attack, more than 100,000 people died, more than the Hiroshima and Nagasaki attacks combined. Imperialism died a hard death.

On the European front, Hitler had developed excellent technology but underestimated his energy needs and fighting abilities in Russia. Hitler's idea of unification under the fascist Nazi regime could place his empire on a par with the United States, his biggest fear and competitor. Specifically, Hitler knew the US had the best manufacturing capabilities in the world. Fortunately for the Allies, Hitler's ego allowed him to populate his staff with yes-men who hid reality and reason from the Fuhrer. Hitler and his staff overestimated their abilities and their capabilities. After all, they were of the historic, superior Aryan race.

World War II

The aftermath of World War I and the enforcement of the Treaty of Versailles initiated a resentful attitude in the German populace toward the Allies. While losing some of its territory, Germany had reparations and limits on the size of defensive military, a further thorn toward national disgruntlement. As the 1920's and 1930's passed after World War I, the situation in Germany worsened with the advent of the Great Depression. The rise of Nazism—greatly assisted by Adolph Hitler's chancellorship—increased anti-German sentiment, while the German antagonistic attitude on their international treatment exacerbated the political situation.

During the same period and before, Japanese leaders had developed a taste for domination over the Pacific arena and Asia, especially since Japanese energy sources were imported. By 1937, Japan had sufficient avarice to sustain a war with the Republic of China, a nation many times the size of Japan. In light of the succeeding attack on Pearl Harbor and the US, Japan's aggressiveness and will for world power was of a magnitude without measure. Rational thinking had evaded Japanese leadership, and these same leaders convinced many of their countrymen to follow an aggressive path to world power. From the late 1800's to 1940, Japan and Germany's political relations were off and on. In the late 1930's, Germany wanted Japan to attack Russia, but logistically Japan was not prepared for such an operation. Later, the overconfidence of Hitler and a shortage of logistics for the prolonged Russian operation dealt a defeating blow to the German army.

As Hitler climbed the German ladder of leadership by becoming the Chancellor of Germany, he and his overzealous staff had not considered their insatiable need for energy supplies in their world conquest. The takeover of Poland in September 1939 was clandestine. Hitler was still deceived by the potential oil production of Poland. The Axis power's avarice was years ahead of their energy capabilities when aligned with their thoughts of world domination.

On the other hand, Japan was aware for years before the war of its limited energy supplies, since almost all oil was imported, much of it from the US. Japan's eye for world conquest, coupled with its lack of domestic oil resources, caused Japanese leaders to invade French Indonesia in September 1940. Primarily, the quest was for oil, but rubber, copper, and minerals were also in great demand in the Japanese Empire. This aggressive, land grabbing move caused the US to pass the Export Control Act to further cut oil and iron exports to Japan. Simultaneously, Germany had conquered a majority of Europe, and Great Britain was on a list for defeat, barring help from the Allies and the US. Japan viewed the western powers as weak, indicating any weaknesses must be acted upon for advantages. Under the perceived condition, Japan formed a closer alliance with Germany. Driven by a mutual shortage of raw materials, especially oil, Japan, Germany, and Italy formed the Tripartite Pact in September 1940. With the pact, the countries would respect each other's leadership and support members if attacked, perceivably by the US.

September 1941 saw Japan expand its presence in Indochina, and such actions provoked Western Powers. US placed a complete oil embargo on Japan, forcing Japanese leaders to capitulate or seize more territory in Indochina. Japanese leadership saw the embargo as an act of war. A state of war was not declared and capitulation and a return to peace in the homeland were still options. However Japanese imperialists' thinking was not to suffer humiliation at the hands of the US.

In the long history of the Japanese Empire that included the Korean Peninsula, the society was a sovereign, self-serving nation, but the desire for power and expansion became powerful psychological factors in which the leaders indulged. The leaders could only see themselves in the glory of power from the world they might create. Some admirals saw the flaw in the Japanese imperialists' thought, but prevailing thought demanded an attack on the US, namely Pearl Harbor. As history reveals, the Pearl Harbor attack was devastating—temporarily. The inability of Japanese leadership to see or forecast the recuperative powers of the US, the Allies, and freedom loving peoples of the world led to their downfall.

From a philosophical view, the defeat of Germany began long before World War II. The aftermath of Hitler becoming Chancellor of Germany in 1933 was the abolishment of any form of democracy or revision of his rules, based in part, on racially oriented views. Eventually, these racial views became laws. While complications may arise, a strong democratic government supports a free thinking and acting people. Unfortunately, the people of Germany were not free. The German masses were thinking only of the present, and of the orations given to them by Hitler; their thinking was introverted. However misinformed, Hitler was an excellent orator; one simply should not listen to his falsehoods couched in rhetoric.

The German attack on Russia, which was orchestrated by Hitler, demonstrated a severe lack of forethought and strategic planning. Hitler was certainly surrounded by generals. But, they were handpicked by Hitler, indicating there was a high probability that they were "yes" men and not really of general officer quality. Where were these supposedly knowledgeable generals when Hitler began talking about attacking Russia? To appease Hitler or to feed their outsized egos, they appeared to or went along with Hitler's grandiose plan to attack Russia. The psychological motivation of power offset good judgment and rationale in the German hierarchy.

In the rush for world rule, German leaders had not prepared for an attack on Russia. The use of correct supplies and means of logistics for troops evaded the planners of the German hierarchy and, certainly, Hitler. Light or medium wear clothing proved to be far short of requirements for Russian winters. Winters demanded stringent logistical requirements be adhered to for the operations of trucks, tanks, airplanes, artillery, and infantry weapons. In addition to the lack of foresight on these crucial items, German strategic planners knew little of the food necessary to feed an advancing army in the winter. The German planners lacked good intelligence on Russian land features, and payed little heed to the fact that Russian soldiers would be fighting on their home grounds. Ancillary to this fact, Hitler ran the Russian war from the safety of his homeland.

From the viewpoint of prisoner treatment, soldiers or civilian, there was no justification, excuse, or intellectual reason for the cruelty of the Axis. Both

Japan and Germany were guilty of prisoner mistreatment that went well beyond societal mores, laws, civility, or guidelines of humanity. Punishment implies an action against a prisoner that is corrective. In the case of either Germany or Japan, neither country exercised corrective actions, but each demonstrated unwarranted cruelties and vengeance on humanity. For each country, the transgressions against humanity are without qualification, and their deserved shame should last forever. Repatriations from each country should also last forever. Currently the attitudes of Japan, Germany, and Italy are such that they believe that World War II was not their fault, and that these countries should not be held accountable for their actions. History and time have moved on, but forgiveness must lie in the minds of individuals.

The Japanese Imperialists committed crimes against humanity that are unsurpassed. A review of the Bataan death march is exemplary for unwarranted cruelty to captured, surrendered soldiers. Records indicate between seven and eight million civilians died under Japanese occupation of China. Within the scope of human societal mores, such a death rate is beyond civilized morality and law. The examples of Japanese cruelty are numerous, but the atrocity of the Nanking Massacre where several hundred thousand civilians were raped and murdered requires emphasis for unnecessary Japanese inflicted cruelty. Why in this same scenario did the Japanese Imperial Army find it necessary to bury Chinese civilians alive? It is impossible to follow any reason justifying such action. A psychological need for merciless cruelty satisfying self-serving egos comes the closest to an explanation. How and why did this need for cruelty arise? Imperialist leadership thinking went awry with greed, an insatiable quest for power, and lust for a larger Empire. Today, Japan would love to deny its past cruelties and overwhelming desire for power, but the facts of history are quite plain. Japan must live with the burden of historical shame, which it cannot deny.

To say the least, Nazi hierarchy and good Third Reich followers possessed thought or reasoning well outside normal human parameters. As a single source of hype, psychological motivation, and utopian promises—Hitler was the answer. His excellent, driving oratory was equaled by a few, but his rhetoric was against the laws of humanity, decency, and society. The German masses were mesmerized by Hitler's by Hitler's Pied Piper promises of Aryan rule over a perfect world. The non-thinking masses fell for the lies and placed themselves under Hitler's control. As history teaches, insidious control over the masses has been exercised in previous periods of monarchical rule. Hitler understood this concept. Without doubt, Hitler, the Nazi party, and the German people were responsible for the Holocaust and the murder of at least six million Jews. Nearly seven million Polish citizens and others perceived as undesirable were also killed—for no known humanitarian reason.

According to Hitler's philosophy, the Aryan race must reign supreme, and to accomplish and maintain the pure race, other races and undesirables must be eliminated. Hitler needed like thinking men and thugs to help his long term dream materialize. These men included generals, doctors, and scientists that postwar investigation revealed as questionable in their professions. From the early 1930's to the close of World War II, preliminary information had professed 7,000 prisoner camps, but later, postwar data contained a list of 20,000 camps—unbelievable. These camps determined the destiny of prisoners—often times death. The Nazis were trying to play God.

The cost of building, manning, and maintaining this many camps was an astronomical figure, but Hitler's henchmen had plundered and stolen from many countries to support his avarice for power and world empire. Of course, many of the camps were primarily operated for prisoner extermination with the secondary and tertiary purposes of slave labor and medical experimentation. Mass extermination of Jews with utilization of slave labor were operations which followed Hitler's dream of supremacy. It is not clear if the extensive medical experimentation was anything more than Nazi doctors' camouflaged sadistic behavior and displays of power. Details of the myriad of medical experiments of experimental procedures defy even a morbid description, but the operations and procedures were only limited by imaginations—especially of the attending "doctors."

For Hitler, the driving motives were to assert superiority of the Aryans. The concept of the Aryan race comes from some history and research of the Tiwanaku ruins near La Paz, Bolivia. The original research emanated from Belisario Diaz Romero, an MD who studied archaeology among many other disciplines. Diaz rejected Christian creationism but espoused natural selection. He was aware of Darwin's theory of common origins but believed in pologenic, or multiple genes, for humans. By his reasoning, the human species was comprised of white, yellow, and black constituents. More refined, was the white Homo Atlanticus race, an ancient Aryan race that migrated from Atlantis. According to Diaz, the Aryans migrated to South America about two hundred million years ago. The disciplines of paleontology, archaeology, anthropology, and geology offer no substantiation for Diaz's theory. With no corroborative data, fabrication of convenience belies any resemblance to the truth. Most anthropologists and paleontologists would agree that five million years for all of mankind's history is a good stretch.

Diaz had no substantive research or data; his conclusions of Aryan migration from Atlantis are unwarranted. Subsequent paleontological data lent some idea of existent conditions two hundred million years ago. Firstly, no human species existed or would for over 195 million years—even at a stretch. This time in world history marked the end of the Triassic Period and the beginning of the Jurassic Period. If humans existed—which they did not—they were standing next to various species of dinosaurs.

Edmund Kiss, a Nazi officer, was enthralled with the ideas of Arthur Posnansky, who was obsessed with studies on Tiwanaku. With little doubt, Posnansky was aware of Diaz's studies and theories. Kiss was the prophet for the Nazis and the Aryan race. The teaching of the Nazi-Aryan-Atlantis connection was fundamental to German superiority, and Kiss delivered the message. From Diaz to the followers of Posnansky and Kiss, it is a clear obsession and usurpation of science backed by biased research. These Nazi quasi-scientists were not true researchers, as they used science as a tool to further their preconceived notion of Aryan supremacy. In their endeavors, true science was nowhere to be found.

When looking at the cadre of MDs and PhDs on Hitler's medical experimentation staff, exemplary credentials are prevalent. However, a scrutiny of such doctors as Clauberg, Schumann, Mengele, Kremer, and others elicits a question of academic rigor. How does a doctor agree with an Aryan race history that dates to the beginning of the Jurassic Period? For findings of medical experimentation, bias in results and the aspirations of political advancement must be considered. In civilized societies with governments operating under democratic principles, the acceptance of findings and data following a regimen of torture are not admissible to any academic endeavor or medical board. For all the torture and pain suffered by Nazi prisoners, the findings are worthless. Over the years various medical studies have wanted to consider the data.

At any time, it is difficult to believe Homo sapiens, which followed the decrees of Nazi fascism, were in a working dilemma for engineering more Jewish deaths in less time. The Aryan concept for superior Nazi humans had flawed research influenced by supposedly credible German scientists. The Aryan-Nazi scientists were adjusting facts to support their forgone conclusions of a superior Aryan race. Backed by the psyche affected by the orations of Hitler and the wholehearted support of henchmen as Himmler and Goring, the German masses were mesmerized and were duped into following. Anyone not of Aryan background was considered inferior. Largely, the followers of the Jewish faith fell into undesirable categories. It seems some Aryan- Nazis thought in terms of a Jewish race, rather than followers of a religious faith. Ignorance reinforced by high ranking- knowledgeable- Aryans was difficult to dissuade in heated political environments, such as Germany in the 30's and 40's. Hitler was a catalyst to ongoing scientific and political ignorance. For those times, a dreadful fate existed for someone to think above the cesspool of quasi-Aryan knowledge, since fear created by the SS and other cadres of Nazi specialists would threaten death or imprisonment to an undesirable in one of the many death camps run by the psychotically deranged Nazi leaders.

When considering human history to the end of World War II, we must acknowledge that world, political, and social changes were indeed made by Homo sapiens. Although modern cultures tend to place humans into various races, we are all Homo sapiens and belong to the same species that made a myriad of culturally malevolent decisions and mistakes. Humans have also made good decisions and have advanced technology and, hopefully, humankind's long-range thinking capabilities will improve the future.

The Holocaust

Over the years since the end of World War II, some naysayers have been quite vocal about the Holocaust's nonexistence. It is difficult to understand the denial of the Holocaust and view countless testimonies of survivors and physical evidence, although the Nazis tried to their very best to hide, transfer, or destroy evidence as the Allies approached at the close of World War II.

With such a crime in progress, 1933 to 1945, the Nazis avoided nearly all written communication about Jewish work, concentration camps, or death camps. Verbal orders were given when the SS or selected Nazis were to enslave or eliminate Jews. Even verbal orders were encoded, reducing assignment evidence to concentration camps or pogroms for massive numbers of Jews. Still, some ardent Nazi leaders knew there would be eventual reparations for the Nazi atrocities against the Jews.

Primarily, the Nazi crimes were conceived from notions in Hitler's book *Mein Kampf,* but the quasi-paleontological data emanates from Aryans migrating from Atlantis to Tiwanku, Bolivia in years before humans existed—what magical madness. For the Jews, the Nazis could not differentiate between religion and race, which places the Aryans on a lower level of intellectual capabilities. Hitler was not from a 100 percent Aryan background, but from his point of view, that fact was not important.

For the Nazis, the camps progressed from detention, work, to death. Some camps performed all functions. As early as 1923, Hitler harbored the idea all Jews should be eliminated, but the Holocaust started in 1933 with Heinrich

Himmler and the SS taking control of concentration camps in the period of 1933-34. By the end of World War II in 1945, the number of undesirables, Jews, and other Nazi perceived misfits was well over 700,000. Word of the Holocaust, pogroms, and death camps had reached Great Britain and the US by 1944 or earlier. To counter the horrendous situation, the Nazis had the International Red Cross, or IRC, visit the Theresiensladt ghetto in Bohemia—now the Czech Republic. The IRC visit occurred only after the Nazis temporarily transformed the ghetto into a homey place. After a perfunctory, six-hour visit by three IRC officials, prisoners were whisked to other camps. Strangely enough, the IRC helped Nazi war criminals escape to Argentina before and after the close of World War II.

The histories of the Holocaust will probably never be complete, since the Nazis, fearing the just and due prosecution of the Allies, expended much effort destroying Holocaust evidence and condemned prisoners to the "final solution"—death. To Hitler's delight, the Warsaw Ghetto was established around November 1940. Except for those who were captives in the Warsaw Ghetto, no person can fully describe or understand the societal laws, abysmal living conditions, or human atrocities committed by the Nazis—as directed by Heinrich Himmler. The predominantly Jewish population and other undesirables or non-Aryans suffered the mental anguish of possible selection for a train ride to Treblinka or other extermination camps daily. A part of the mental suffering came from a list of known or unknown transgressions; one did not know which offense of the day might be the cause of an individual or a family getting a train trip that day. For the Jews and other undesirables then, there was only day to day living, possibly hour to hour. The longer range planning of normal life could not to be considered.

Although Warsaw was a fairly large capital city in Poland, the Jews were confined to under two square miles within the city. But the unimaginable is that over 400,000 people were squeezed into this living space. Very shortly after the beginning of the Warsaw Ghetto, all living conditions deteriorated to far below a civilized person's concept of existence. The Nazis wanted the badly deteriorated condition to foster more deaths, and the deaths by typhus, starvation, or medical problems did occur. According to Himmler and ranking Nazis, the Jewish death rate at Warsaw was nearly high enough; so, over a two-month period in mid-1942, over 250,000 Warsaw Ghetto residents were shipped to death camp Treblinka.

During the operation of Warsaw, over 300,000 Jews were transported to camps for work or the "final solution"—death.

An easy method for the Nazis to increase the death rate in the Warsaw Ghetto was by disease caused in part by starvation. Employment within the Ghetto was nearly impossible; procurement of food through monetary means was nonexistent. Small children smuggled for families as best they could. Jews in the Warsaw Ghetto were rationed a little over 180 calories per day, while calories for German citizens were over 2,600 calories per day. As per Nazi unwritten decree, death by disease caused starvation was the plan for Jews. From Warsaw to the death camps, the pervasive Nazi-Himmler attitude was: why feed, comfort, or medicate a group of undesirables we are going to exterminate?

It's nearly impossible to imagine how Jewish families of four or multiple sequestered families managed subsistence on a little more than 180 calories per day per person. For continued living, the ingenuity for food procurement had to match that of engineers building complex machines. Procurement was limited by the obstacles set in place by the Nazis and imaginations of the severely oppressed Jewish families.

While the Nazis had many work, concentration, and death camps, Treblinka played a major role in the death of Holocaust victims due to its proximity and convenience of the Warsaw Ghetto. The fears of the prisoners of the Warsaw Ghetto were justified in that the commandants of Treblinka would easily replace workers with whomever they required from Warsaw. The replacements were completely at the whim of the current resident Treblinka Commandant. It was more than likely that a train trip to Treblinka would inevitably and assuredly lead to death.

The Treblinka extermination camp was located near the village of Treblinka, northeast of Warsaw and operated from July 1942 to October 1943 as part of Operation Reinhard. With its close proximity to Warsaw, Treblinka was prime to the Nazi "final solution"—Jews' deaths. The number of deaths is not known, but most academics give a variance of between 700,000 to 900,000 deaths at Treblinka. However, several credible authors give the death number at 1.2 million to 1.58 million, and author Vasily Grossman in his book, *The Hell of Treblinka*, gives a death number of three million. We will never know the exact number, since the Nazis shamefully and fearfully destroyed all records as the Allies and Soviets approached. Understandably, civilized societies loathe these nearly astronomical death numbers, but considering the Nazi avarice for Jewish deaths and their loathing of the Jews—while boasting of the superiority of the Aryan race—it tends to bias judgment toward higher numbers.

For functionality, Treblinka was divided into two camps, Treblinka I and Treblinka II. Treblinka I was a forced labor camp, where many people died of exhaustion, starvation, and disease—usually typhus. Treblinka I's workers labored in the irrigation areas of the surrounding forest, cut wood for the crematoria, and worked the gravel pits.

Treblinka II was truly an extermination camp and a prize for the Nazis. Workers concentrated on burying victims from the crematoria, and, later, digging up the victims for burning and hiding them before the approaching Soviets. Initially, the gassing of victims was from one or two tank engine exhausts, but a visit to Treblinka II from the Auschwitz Commandant, Rudolf Hoss, instigated the use of the efficient chemical Zyklon B, which is a lethal cyanide gas. With the increased efficiency of Zyklon B, the killing rate could increase to as much as 25,000 people a day, but the maximum rate of deaths per day was closer to 15,000.

Prisoner maintenance in Treblinka was nonexistent from the Nazis and, in fact, Himmler wanted more deaths while prisoners lived and worked. On one occasion, a prisoner train arrived with 90 percent of the prisoners dead. The deaths were caused by suffocation, starvation, dehydration, and disease, which was without medical care. The attrition started before the death trains arrived, and the conditions were so miserable in Treblinka I or II that twenty suicides may have occurred per night.

The list of atrocities and inhuman activities at Treblinka, Belzec, Sobibor, and Auschwitz, and thousands of other concentration camps cannot reasonably be numbered. As the world progresses in history, we tend to disregard or forget about the happenings and horrible situations of World War II. Time tends to dim the events and learning, but the atrocities, suffering, deaths, and sorrows of people living the time were real, especially to the Holocaust victims. In our schools, little is spoken of the Holocaust, and, yet, the imprisonment and killing of six million European Jews or perceived undesirables was a major, world-class atrocity to humans by humans. In history World War II constitutes a recent event. But have we learned from it? With no knowledge of paleontology, anthropology, and ancient history, Hitler and the Nazis declared the superior Aryan race was formative to Germanic culture and Germany. Research and science does not confirm this abhorrent, ill-conceived fascist idea. With little doubt, Hitler's orations had supplied the initial enthusiasm for the goal of all—something for nothing—German society. Germany has started two world wars. Have we forgotten? Germany still bears some watching—how unpopular and politically incorrect. Are we too quick rushing to normalcy and assuming a complacent posture?

AUSCHWITZ CONCENTRATION CAMP

The Nazis, in the pursuit of Empire and Aryan supremacy, found the need for many concentration camps to utilize or eradicate their perceived idea of inferior races. Their pearl of a concentration camp was Auschwitz, which was comprised of three camps, Auschwitz I, Auschwitz II-Birkenau, and Auschwitz III-Monorwitz. While Auschwitz was the original camp, Birkenau also served as an extermination camp and Monowitz was a labor camp.

In pursuit of the Aryan and supremacy idea perpetuated by Hitler, the Nazis spared no efforts, or the efforts of Jewish prisoners, to build extermination camps for the immediate or future demise of Jews and other non-Aryan undesirables. The estimate of deaths at Auschwitz is probably 1.5 million, while the Russians placed the number around four million, which historians believe to be high. The means of death for railroad prison car Jewish prisoner arrivals were only limited by the Nazi mastermind's imaginations. Arriving prisoners were initially separated into two groups, depending on the Nazi officer's or doctor's judgment on their abilities to work, unless they had physical attributes or maladies warranting immediate death. Mass deaths by gassing of carbon monoxide or the later method of Zyklon B were the most efficient methods. The prisoners destined for extermination were led to believe showers were in store. After a mass gassing of Zyklon B, bodies were moved to one of five crematoria in Birkenau for cremation. The record number of deaths per day has been estimated at 20,000.

The methods of death were not limited to the shower chambers. Killing with phenol injections, intravenous or direct to the heart, proved to be quick and efficient. Joseph Mengele was reported to have used the method, but while Mengele was the chief Doctor, Friedrich Entress proved himself to be the prima donna of phenol injection death procedures. Apparently endowed with few personable attributes, his avarice for Jewish deaths made him the most lethal Doctor in Auschwitz. Certainly the list of phenol injections was not shortened with proponents like Bodman and Grabner who joined with Mengele and Entress. An able, overzealous assistant to Entress was Josef Klehr, whose eagerness to inject phenol and kill was well outside of medical, societal, or parameters of civilized living—in any history period.

The Nazi goal for the Jews was complete annihilation, and any method to this end was considered. Hitler had appointed Heinrich Himmler to establish the Nazi concentration camps, and since Himmler was an excellent organizer, many camps materialized and became quite efficient at killing, including Auschwitz. Himmler's goal was mass, efficient killing, and various methods were used in experimentation. One method was straightforward, cheap, and effective—starvation. If upon arrival by rail, a prisoner was assigned to the right line, temporary life and the Monowitz labor camp was a likely destination. In this camp and others, Nazis had figured three months of labor on minimum rations was the maximum for work and life. Different scenarios simply place prisoners in a ward and let them starve.

Auschwitz was used to starve prisoners, and on a grander scale through Hitler, Himmler, and Herbert Backe, the Hunger Plan was hatched. In addition to starving people in camps, operation Barbarossa, the Nazi invasion of Russia, completely hinged on the German army living off the land. Yet another reason for depriving Auschwitz and other camps of food. The Hunger Plan designed by Backe intended to starve 20 to 30 million Russian people as the German army fed itself and conquered Russia. Incorporated in the Barbarossa plan was diversion of food from native Russians to a starving, freezing, and stymied German army.

The most poignant reason for the failure of Barbarossa was a lack of realistic planning on the parts of Hitler, Himmler, and Backe. The idea of an invading army living off the land in the Russian winter demonstrates a definite lack of knowledge and leadership. What gave these planners and idea several million Russians were going to helplessly stand by and starve? Food diversion from concentration camps was equally preposterous. Thirdly, the Nazi planners had no scope of the distance involved or logistics management. While the army was to self-sustain, resupply of hardware logistics on nearly kaput railways was not reasonable military thinking. Of course, food deprivation for the concentration camps was a norm, but not supplying food to an advancing army is heresy.

The deprivation idea ran deep at the Auschwitz camps. The execution chambers provided were so tightly sealed prisoners could be killed simply by placing them in a chamber and allowing oxygen deprivation take its course. The chambers were dark so that candles could assist in lighting as well as burning oxygen.

In conjunction with starvation were experiments on prisoners who ranged from new arrivals—fairly healthy—to advanced states of emaciation and edema. Prisoner physicians performed the experiments under the supervision of Nazi doctors. The results of the experiments were carefully prepared and published in a scientific report after the war. Apparently the report was the only one published by prisoners. Although published by prisoners studying the long term effects of starvation such as emaciation, edema, and plasma with lowered protein and albumin, the validity of any such study produced in Auschwitz is questionable. Nazi doctors in the early days of concentration camps thought to publish their medical experimentation findings. Considering the cruelties of Auschwitz to include starvation, the findings or conclusions of any Nazi experiment were invalid. In most cases, prisoners were gassed after an experiment, especially those with edema and emaciation.

Later in the war and when the Nazis could plainly see the Allies were winning the war, the Nazis wanted to cover the mass evidence of numberless killings, work camps, medical procedures with no motive for healing, uncountable thefts from prisoners and their resident countries, inhuman experimentations, and cruelties beyond imagination. In the camps, Nazis were busy digging up bodies, burning, and destroying incriminating evidence before the prying eyes of the approaching Allies. Although the Allies were surely and quickly approaching, the Nazis still attempted to destroy camps, gas chambers, cells, torture devices, and medical paraphernalia. However, since the Nazis had spent so much effort and money on their endeavors to establish the supremacy of the German Aryan race and eradicate the Jews and other undesirables of our species, they couldn't begin to destroy their mountains of incriminating evidence.

This begs the question: why did the Nazis feel the need to destroy their edifices and evidence to further humanity? Apparently, there were those in upper echelons of Nazi hierarchy that understood from the beginning of the Aryan philosophy that the reasoning and motives were flawed. The ideas of control, power, and wealth were powerful enough incentives to temporarily overcome reason and wisdom for the Third Reich. In short, some high-ranking Nazis acquired some psychological guilt, and Auschwitz was the tangible, condemning evidence of their selfish, malicious thoughts.

World War II Japanese POW Concentration Camps

World War II history, documentaries, and Japanese POW testimonies lean towards a strong penchant for prisoner mistreatment that included: starvation, lack of medical treatment, lack of medicine, exhaustion, lack of or poor sanitary facilities, lack of or poor living quarters, and generous amounts of torture or disciplinary actions. From the historical stereotype of concentration camps to a myriad of so-called "hell ships" operated by the Japanese, the misleading reasons for Japan to administer unnecessary torture, death, or both for all people representing the Allies were unclear. A fascist Empire driven by war for oil and energy was a national motive. The most obvious reason for excess POW deaths and torture was that the Japanese relished inflicting pain and death to POWs of greater nations. The Japanese did not sign the Geneva Convention of 1929 and made a point of reminding the rest of the world. For the Japanese fascist inflicting pain and death to soldiers of Allied nations was a measure of equality.

Why was the death rate of Japanese held POWs over 27 percent higher than prisoners held by the Germans? There is no rationalization for the preponderance of deaths at the hands of the Japanese POW jailers, but the underlying Japanese theory was that surrendered soldiers, POWs, have no honor and should have died on the battlefield. Therefore, there is no dishonor in dying at the hands of the Japanese but not before a great deal of labor has been extracted.

During World War II the number of concentration camps, jails, prisoner or detention areas was nearly uncountable, and the atrocities committed against prisoners were unimaginable and unquantifiable. With the numbers of prisoner facilities disproportionate to the size of the Japanese Empire, perverse behavior poses as a purpose of imperialistic thinking. The phenomenal number of cruelties inflicted reveals positive feedback for those inflicting the pain. As evidenced, prison maintenance with associated prisoner torture appeared to be primary in Japanese war efforts, rather than the actual activities needed by a world conquering Empire.

The examples of Japanese cruelties are many but the atrocities at the Puerto Princesa prison camp are representative. In short, it contained the Palawan massacre whereby Japanese soldiers and the military police murdered 139 US POWs, following the imperialistic sentiment that POWs should die. In the last day of 1944 to prevent rescue by advancing Allied troops, 150 POWs at Puerto Princesa were herded into air raid shelters that were subsequently set on fire. Prisoners caught escaping were dutifully bayoneted. Any successful escapees were later hunted down and summarily shot. The episode represents one of the countless examples. After World War II, reprisals and punishment for the Japanese Empire could not match the transgressions and operations outside of socially accepted mores the Japanese had committed.

. Looking at the transportation sagas of POWs reveals only sadistic Japanese behavior. Following Japanese custom, most ships were called after a proper name followed by "Maru" – meaning beloved. In September of 1944, the "Junyo Maru" was sailing to Padang, Sumatra with 6,520 POWs aboard. West of Sumatra the 5,065 ton, 405 feet long ship was torpedoed. As usual, Allied submarines had no way of knowing the ship was loaded with POWs. The ship was probably built in 1916 and eventually came under Japanese ownership. As can be imagined, the small ship needed special outfitting to carry over 6,000 prisoners. Of course, bamboo was used for extra flooring and cages to control and house the prisoners. The word deplorable would glorify the filth, stench, disease, hunger, heat, dehydration, and pestilence as conditions for travel. Psychologically, most prisoners had long departed from the aspiration of hope when the "Junyo Maru" was torpedoed. 5,640 POWs lost their lives, and at the time, the "Junyo Maru" sinking was the world's worst sea disaster.

Other Japanese ships were utilized in a similar manner. Many ships reached their destinations only to have their prisoner cargo work, languish, or die on some Japanese work project. And, of course, other prisoner ships were sunk, even to have Japanese sailor's machine-gunning swimming, escaping prisoners. Human mores or psychotic behavior cannot be stretched to cover such actions against humanity. From the beginnings of humanity, World War II can be considered distant, but from the first human civilizations, World War II represents a recent point in history. When motives and aspirations of world dominance and control pervade the minds of greedy leaders such as those of Japan, Italy, and Germany, rules of society disappear quickly. The world must guard against repeated World War II actions and the human thoughts inspiring those actions.

WORLD WAR II U-BOATS

The stories and facts of World War II are without end, and the saga of the North Atlantic War deserves some investigation. There was strife in the North Atlantic from 1939 until Germany's defeat in 1945, and the bulk of the shipping conflicts occurring from 1940 to1943. Since the UK was an island country, it was dependent upon shipping and imports for much of its subsistence. Hitler saw a curtailment, if not a complete shutdown, of shipping as an excellent method for controlling and conquering the UK. Heretofore, submarines as a fundamental tool in warfare were not in hierarchical thinking of the admiralty. But for slowing down or stopping shipping in the North Atlantic—the submarine was the ideal tool. With U-boats in operation, Winston Churchill knew the Battle of the Atlantic would be the dominating factor throughout the war. The battle could not be taken lightly. Since from the middle of 1939 to the middle of 1943, 500 merchant ships, 175 warships, and 783 U-boats had been sunk. In the North Atlantic mayhem, over 36,000 sailors, 36,000 merchant seamen, and approximately 30,000 U-boat sailors died from all aspects of war; such death tolls are unacceptable. Except for the desperation of the situation, admirals and generals turning in such losses should have been relieved of duty.

By September 1939, the European War had begun, but US leaders wanted to play the role of isolationists. For this attitude of denial, the US paid a heavy price, which included humiliation. History has proven US leaders do not want to panic the populace; therefore, it is better not to publicize some major problems or defeats. The people living near the coast of North Carolina could see and testified to the U-boat terror and sinkings offshore during World War II long before the US and Admiralty began convoys and merchant shipping protection. At the time, debris, bodies, and oil slicks were in the sands on east coast shores. Even today, oil sludge can be found below surface sands on some beaches. The so-called "Torpedo Junction," off North Carolina's Outer Banks, saw much U-boat action and caused an estimated 150 million gallons of oil to settle into the sands. Still, neither the State Department, Navy Department, nor Admiral King, saw reasons for blackouts on either US coast. Denial can rise pretty high in the ranks.

While World War II progressed, several submarines had deposited spies and espionage crews along the eastern seaboard to include Canada. For example, June of 1942 saw German spy George Dasch and eight other agents aboard U-202 arrive on the shores of Amagansett, New York. The havoc ordered by Hitler was to destroy Niagara Falls power plants and aluminum manufacturing in Illinois.

To be sure, all the sabotage and spy missions are not known, but U-584 tried to place mines in transportation systems from a landing point at Ponte Vedra Beach. Another landing by U-1230 near Hancock Point, Maine tried, unsuccessfully, for more sabotage. Hitler and the Nazis were determined to get the upper hand on the North American seaboard, and to this end in 1943, U-537 landed at Martin Bay, Labrador for installation of a remote meteorological station. U-boats could have had reports of current North American weather.

Eventually, there were no safe harbors along the North Atlantic seaboard, and because of oil shipping, the Gulf of Mexico was occupied by U-boats, with one being sunk. Along and into the Saint Lawrence Seaway, U-boats were brave enough to ply their trade, even considering the shallow waters. From the time the Germans declared war in 1939, the United Kingdom incurred a need for imported supplies. A disaster was occurring to North Atlantic shipping, even with the loan of outdated destroyers from the US to Great Britain, but by 1941 US sentiment was growing against Germany. US merchant ships and those of other countries were being harvested by German U-boats, while the US had done little for North Atlantic protection. US leadership prophesied: let Great Britain handle the havoc—it's their problem. For an embarrassing period of time, US leadership did not realize the magnitude of the shipping crisis, nor the worldwide problems growing in the maintenance of international sovereignty.

Finally, the US leadership and the Admiralty addressed the monstrous North Atlantic shipping problem, but the Japanese attack on Pearl Harbor clearly brought the US survival problem home. By June 1941, the British were providing merchant Marine escorts across the entire Atlantic, while President Roosevelt's Pan-American Security Zone allowed American warships to escort vessels as far as Iceland. Mid-1941 saw more organization of shipping by the US, but the involvement was late.

Fortunately, for Admiral King, US leaders did not fathom the criticality of the North Atlantic shipping situation, or he would have been relieved of his position. A few of the reasons Admiral King abstained from early protection of the eastern seaboard and was less than enthusiastic about warship escorts on North Atlantic convoys were: Coast Guard movements were predictable, the US Navy was short on proper escort vessels, and interservice communication and cooperation were poor—partly his responsibility. Admiral King did not respect British recommendations, and, apparently, only the President could counter his theory of non-aggression.

During World War II, the predictability of certain actions was a difficult problem. At the outbreak of the European World War II, the US was short of fighting vessels, but the supply of a few US Navy manned fighting ships for North Atlantic convoy escort duty to the UK might have provided more resistance to Germany and the U-boat fleet. It is a reasonable assumption that an earlier increase in war making materials and supplies would have bolstered British resistance to Nazi aggression. The European war might have been possibly shortened, but leadership must have foresight, think internationally, and have reasonable forecasts for consequences.

WORLD WAR II JAPANESE SUBMARINES

Prior to World War II, an estimate of Japan's submarine fleet would have yielded few. Surprisingly to the contrary, Japan held a world-class, technically advanced submarine fleet and in January 1943 was secretly building the I-400 class submarines. Plans for the use of these submarines varied, but they assuredly were intended for increased Japanese fascism in the world. The Japanese Imperial Navy had a majestic plan for, possibly, 21 of these Sentoku 400-foot-long, submarines but, luckily, the Japanese could produce only three. These Tokugata Sensuikan—Sentoku—submarines were a special breed with increased capabilities: 4 x 2, 250-2,400 horsepower diesel engines, sailing capabilities of one and a half times the Earth's size, crew of 144-220, surface speed of nineteen knots, submerged speed of twelve knots, displacement of 6,560 tons, and dive depth of over 300 feet. Fortunately, the initial manufacturing of these I-400 submarines started after the beginning of the war, which hampered production in numbers. No other Navy, including the US, had submarines with these capabilities. Such capabilities did not appear until the nuclear submarines of the 1960's.

Admiral Isoroku Yamamoto instigated the Sentoku, I-400 submarines and had some world changing, destructive plans in mind. Purposefully, the I-400 submarines could sail from Japan to the US east coast and return. Specifically, New York City, Washington DC, Boston, Philadelphia, and other large cities were targets. It does not seem reasonable that a submarine could target cities that were not clearly on the seaboard. However, when Admiral Yamamoto ordered the submarines, they were to have the capability to launch and recover the Auchi M6A Seiran floatplane aircraft. No other submarine had this capability. The Seiran could carry an 1800-pound bomb at 295 miles per hour. Of course, for the I-400 to launch aircraft, it needed a watertight hanger to hold three aircraft, and places to store the aircraft pontoons, which they used. The I-400's had a catapult launch system. While the Seiran aircraft could be launched without pontoons, the returning aircraft would be relegated to a ditching upon return. Kamikaze missions were considered. From a mission to the US east coast or the west coast, an I-400 submarine could return to Japan unrefueled.

Admiral Yamamoto had formed some insidious plans for the US using the I-400 submarines. First, an attack on west or east coast cities with the Seiran aircraft would be devastating and demoralizing to US citizens. But the infliction was planned to be far more devastating than a single attack. Japan had some history with experimentation with biological warfare. A biological bomb delivered by a Seiran aircraft launched from an I-400 submarine near either coast at night could rain bacteria or fleas infected with cholera, typhoid, anthrax, or bubonic plague to millions of US coastal citizens. Japan's biological warfare attacks on China killed many thousands of Chinese over a period of years, affirming the killing technique. But nearing the end of the war with the Allies winning, the Japanese General Staff opted against the use of biological warfare, as the effects could spread to all humankind.

Closer to the end of World War II, the Japanese considered using the capabilities of the I-400 against the Panama Canal, since there were reports of the Allies not guarding it well. The reasons for attacking were obvious, since the US and Great Britain were supplied via the Panama Canal. If either the locks at Miraflores on the Pacific side or the Gatun locks on the Atlantic side were damaged or destroyed, requirements for repairs would take many months, and Allied war efforts, including convoys, would be slowed against Japan and Germany. Draining Gatun Lake could close the canal for a very long time. During this time of the Panama Canal attack consideration, the Allies were attacking Okinawa and pushing towards the Japanese homeland, forcing the Japanese Admiralty to think about homeland protection. In a good sense, the attack on the Panama Canal was canceled.

It is interesting to note that the Allies, and the US especially, dodged several catastrophes that were called off by the Japanese, and if they had proceeded with their plans, there would have been a certainty of suffering from biological warfare. To a lesser degree, the canceled attack on the Panama Canal spared the US some suffering. The US and Allies were completely blindsided by the I-400 submarines and their inordinate capabilities. For submarines, the Allies had nothing remotely close. Before the war, the US and the Allies had sorely underestimated the motivation and capabilities of the Japanese Empire. Only with the American private sector—with its mass manufacturing capabilities—coupled with our highly trained and motivated military forces were we able to overcome the fascist quest for world power.

Messerschmitt ME-262

For the time and the world's level of technology, the Messerschmitt ME-262, a German jet aircraft, was far ahead of the US, Soviet, or Allied aircraft. The designs for the ME-262 Schwalbe jet fighter had begun in the late 1930's before World War II. Germany produced about 1,400 ME-262s, but fewer than 200 were operational. Contributing to this low number were the mass bombings of the motherland by the Allies. The overpowering production of US private manufacturers, turned armament and aircraft manufacturers, was far greater than the fascists could match. In World War II, massive manufacturing overcame some advanced technologies of the Germans and Japanese. But if Hitler had the raw materials, utilized his generals more, planned better oil reserves, and spent far less efforts and monies on concentration camps, he could have utilized the ME-262 and the ME-163B rocket fighter programs to greater advantages.

Unfortunately, for the Germans, the ME-262 came into the war too late. Regardless of its production tardiness, the ME-262 had capabilities and innovations many years ahead of Allied fighters, and, surprisingly, over jet fighters in years to follow. In its combat role, the ME-262 was 100 miles per hour faster than the P-51 propeller driven fighter. In cruise, the ME-262 was about 115 to 120 miles per hour faster. Aviation technology did not install leading edge slats on jet aircraft until after Allied experimentation on captured ME-262 jet fighters, which already had fully functional leading edge slats. The slats lowered the stalling speed to a range of 99 to 106 miles per hour, depending on load. As the ME-262 had a high wing loading, the slats helped it with tight turns and high-speed fighter maneuvers.

After the war, comparative tests were made against the P-80, which was developed after the ME-262. In these tests lasting nearly five hours of flight time, the ME-262 was better at acceleration and speed with similar climb performance to the P-80. For flight time, the ME-262 had a maximum of one and a half hours. Interestingly, in the mid-1930's, Adolph Busemann had proposed wing sweeps for the HGII and HGIII versions of the ME-262 of thirty-five degrees and forty-five degrees respectively. Only imagination can tell the results had these jets been manufactured and used against the Allies. The original ME-262 design had a wing sweep of eighteen and half degrees for center of gravity purposes, but the F-86 had the wing sweep of the proposed ME-262 HGII, thirty-five degrees, and from history and performance, the F-86 was one of the best fighters of the time.

A few two-seat trainer versions of the ME-262 were manufactured, but they evolved into night fighters with a Neptun radar system and antler antenna array. As can be seen, these aircraft were forerunners of two-seat, multiple duty fighters to come. Fortunately, Hitler did not understand the capabilities of these aircraft, nor did his generals advise him otherwise.

No one can say for sure what variables would have changed the course of the war in time. It is clear the Allies would have spent whatever efforts were required for victory. Against the Allies the fascists had some formidable weapons, but they were not timely in their implementation. In denying raw materials and energy, the Allies can take credit. Greater difficulties and hardships would certainly have materialized if the Japanese had utilized the I-400 submarines as originally planned, and, conjunctively, if Hitler had fully exploited oil reserves and preplanned manufacture of more ME-262s, the conquest of Germany could have lasted much longer or been in doubt.

From the close of World War I until World War II, the US and Allied intelligence agencies were maybe too trusting of the Treaty of Versailles and too complacent with the now "peaceful" nations, like Germany. Since World War I was the war to end all wars, the US was far too quick to downsize the military with a special emphasis on the Navy. Great Britain was quick to follow suit. Regardless of the unwarranted demilitarization, the Allies should have maintained an excellent intelligence network in Europe and on the Pacific Rim, but they did not. The middle and late 1930's produced some fascist warlike symptoms; yet, the US isolationist's point of view kept us in denial. The increased manufacture of Japanese submarines to include the colossal I-400 class, and the advent of a German jet fighter, ME-262, should not have been surprises, but the US and the Allies were caught without a clue to their existence.

Peenemunde

From years before World War II, amateur German rocketeers established a precursor to the Nazi rocket program developed at Peenemunde—the German experimental and developmental station near the Baltic Sea. Many US citizens may have difficulty making a connection between Peenemunde, a Nazi scientific and experimental facility, and the well-developed, space exploratory agency, NASA, in the US. In the post war years, for the US to utilize some of the brilliant scientists such as Wernher von Braun from Peenemunde, US intelligence recruited them to the Operation Overcast, which subsequently was better known as Operation Paperclip. Germany and the Nazis had concentrated its scientists in aerodynamic war making efforts, especially toward the A-4 or V-2 rocket in Peenemunde, which rapidly accelerated progress. In attendance were top scientists from all fields.

Primary at Peenemunde was the production of the A-4 / V-2 rockets, which could carry 2,000 pounds of explosives from Germany to London in a few minutes. The bi-fuel rocket used liquid oxygen and alcohol for its rocket engine. There is no doubt that the V-2 was the world's first guided ballistic missile and the product of Wernher von Braun. Eventually, the V-2 was exotic and reliable in all respects. Thrust from the fuels lasted for about one minute, while a gyroscopic guidance system controlled tabs on the fins. Fortunately, the V-2 suffered some inconsistencies on initial alignment, electrical variations, or premature engine shutdowns. At apogee, the V-2 reached about fifty-two miles, and upon descent it went supersonic. Depending upon the target, the V- 2 could reach its target in just a few minutes.

Hitler and his staff wanted the range of the V-2 increased to intercontinental; specifically, Germany wanted to target New York and Washington DC. The rocket team was working on the problem. Possibly, another engine for the rocket could help. Also at Peenemunde scientists were working on a nuclear weapon, and if developed, the weapon would be loaded onto an Intercontinental V-2 with a target of New York or Washington DC. Luckily, for the US and the Allies, the Peenemunde scientists were running out of time, as allied armies were closing in on Germany.

The possibility of a different ending to World War II was close. A major factor for the US was its vastly superior speed and manufacturing of war materials: tanks, artillery, munitions, aircraft, and equipment for military personnel. However, several scenarios existed for the Nazi production of an atomic weapon. Nuclear power was under consideration by the Nazis, whereby deuterium, heavy water, had been produced in larger quantities at Norsk Hydro in Norway. This was a separate operation from the nuclear weapon experimentation in other parts of Germany. By early 1943 two scientists, doctors Martin and Kuhn, had developed a sophisticated centrifuge, which was something like 3,000 percent more efficient at enriching U235 than any method used by Manhattan Project scientists in the US. Krupp Industries in Hamburg harbored the German enrichment centrifuge, but other scientists in Germany were working on nuclear projects as well.

By integrating the improved V-2 rocket with enough range for US east coast targets to a nuclear weapon, the Nazis could project mayhem on the central cities of the Allies. From postwar findings, it seemed apparent that Peenemunde scientists were further ahead on the V-2 rocket than nuclear weapon scientists were on their project. But as details filtered into intelligence agencies, testimonies of reliable sources reported nuclear test fireballs on the Aisle of Rugen in the Baltic Sea, which could be seen from Peenemunde. As far as completely fissionable reactions, questions have arisen, but a nuclear reaction—to some degree—did occur. Additional evidence revealed 700 prisoners died—many with hideous burns—during the tests.

The science and advancements of the Third Reich were tremendous and superior to US and Allied weapon technology. The V-2 and the ME-262 jet fighter are prime examples. Nazi leaders lacked the commensurate abilities of their scientists. Hitler failed to push the V-2 rocket program to the fullest and thought the ME-262 jet fighter should be used as a bomber. Scientists did not hardly push the nuclear weapon development; perhaps lack of insight provided the deficit in motivation. In any case, Nazi leadership had not provided for future materials or oil energy for empire making.

In 1945, many Germans, Nazis, and Peenemunde scientists could clearly see a dismal fate if they surrendered to the Soviets. As the chief technician, von Braun and his staff opted for a try at surrender to the Americans. Eventually, through a process of a paperwork application, von Braun and many Nazi scientists were recruited by Operation Paperclip, which obscured the scientists' Nazi backgrounds. Certainly, some of the scientists were deeply enthralled with their work and were not devout Third Reich enthusiasts. Their allegiance was to science and secondarily to whatever political regime allowed them their discipline. Others, however, were true Nazis, but became aware of their misguided fascist faith. Of course, all were guilty of Nazism. This cadre of scientists, including Wernher von Braun, were needed in the US aerospace program, essentially because of their success in the Peenemunde V-2 rocket program and valuable experience.

In June of 1945, von Braun and Peenemunde scientists were transferred to America, and the US Joint Intelligence Objectives Agency helped the former Nazis to receive identity papers and clearance to work in the US. Various intelligence agencies interrogated the scientists before they transferred to Fort Bliss, where they found new allegiances in training Army personnel on the issues of rocketry and guided missiles. Eventually, von Braun led the Army's Redstone Arsenal team, which produced the initial, successful nuclear ballistic missile trials. Von Braun's intelligence and rocketry expertise placed him as the director of operations division of the Army Ballistic Missile Agency, and from that division, the Jupiter-C would place the satellite, Explorer One in orbit. With these initial Redstone rocket advances, von Braun had placed the US on a par with Russian orbital endeavors. More successes in rocketry were to follow, but it can easily be seen that progress and meaningful direction has not always come from US citizens or indigenous people. Sometimes progress or advantage comes from recent immigrants who see opportunities in countries offering freedom.

As nations, especially the US and Russia, competed in the space race, von Braun supplied ideas and specifics for traveling to the moon and Mars. Von

Braun was not bantering about grandiose plans; he had already completed engineering calculations and described the parameters to follow. While the projects were described as large, he utilized minimum energy orbits for Earth-Mars trips. His project ideas were workable, but they were a few years ahead of US space conquest thinking. In the late 50's with the Soviets successfully launching an orbiting Sputnik, US sentiment placed its space program in a secondary position—capitalistic humiliation. Basically, to emerge from the national embarrassment, authorities chose von Braun and his ex-Peenemunde, German scientists to create a means and vehicle to introduce productive US orbital launches and bring the American space program out of the woods. In retrospect, von Braun had proposed such a project in 1954, but the US was probably a little too close to the Korean War debacle.

Following soon after the Soviet Sputnik launch, NASA was established in July of 1958, and, shortly, thereafter Redstone successes were noted. Reason mandated von Braun be NASA director, and from mid-1960 to early 1970, he functioned as the first director. The continuance of the Saturn program was a stipulation of von Braun, and the Marshall Center was the agency capable of getting the Saturn rockets to carry heavy payloads either into Earth orbit or space. Von Braun's special desire was to further mankind's knowledge with a mission to Mars. Had the US and political sentiments been more sympathetic to von Braun's aspirations for manned missions in the 50's, the reality may have occurred in the 60's. However, hindsight and speculation prove nothing, but a forward-looking philosophy can become a reality.

Other visions of von Braun's included a heavy duty, earth orbiting space station from which many missions could be launched. Secondarily, the space station could be used as a geodetic survey for Earth, an observatory, and a refuel point for deeper space exploration. However, a lunar orbit rendezvous idea did materialize, and it became the Apollo program. To help develop the program, von Braun utilized an old Peenemunde teammate, Kurt H. Debus. He was the first director of the Kennedy Space Center. As history has provided, the Apollo 11, eight-day mission placed men on the moon in 1969, and for the time, the mission seemed nearly impossible. From NASA, von Braun, and other brilliant scientists, Saturn V rockets reliably placed six astronaut teams on the moon. There were times in the beginning Apollo stages that some people thought the program was behind or being delayed due to von Braun's caution toward safety, and to the safety side, von Braun engineered redundancy above the required strength. The Apollo engineering and excellent safety record can withstand scrutiny, and compared to the Soviet and other national programs, the US safety record was superlative.

In 1970, von Braun became the associate administrator for planning at NASA in Washington DC but elected to retire in 1972. Perhaps, as a scientist, von Braun's long-range vision for the US space program was incompatible with the illogical political ambitions at the time, but included in his futuristic thoughts or ideas was a Space Camp for youth to experience science and space innovations created by necessity.

In retrospect of Wernher von Braun's contributions to science and the US Space Program, a bit of irony exists in that the US should learn so much from a former Peenemunde Nazi and enemy. Truly von Braun was not so interested in being an enemy but in being a contributor to mankind and science. Certainly, after his political allegiance and realignment to US thinking, he had some say in the direction of the world.

GREAT BRITAIN AND WINSTON CHURCHILL

Winston Churchill was an enigma of a British statesmen, having served in many government capacities. Twice he was elected Prime Minister, and during his first term in World War II, he was concurrently the Minister of Defense—powerful positions. In British history, there were no more powerful prime ministers. Churchill's earlier days in Britain's crusade for worldwide rule had proven him as the best and bravest of soldiers. He had gained the award of Britain's Victoria Cross, but did not receive it, since he was a civilian at the time. His eloquence and orations in times of crisis were world-renown and preserved for history students. In the bleak days of World War II, when Hitler was bombing England and London, the only entity holding the country together was Churchill and his rousing speeches of nationalism, sovereignty and never surrendering. The progression of German V-2 rockets and bombers made England's survival a matter of question—except to Churchill. His inspirational, impassioned words and radio speeches pulled Londoners from disparity and probable capitulation. Clearly, Britain was in a materials and armament conundrum for sovereign longevity, as witnessed by the Land Lease Program which was supported by ship convoys from the US and Canada with vital war supplies. The catalyst for these nation saving actions was Winston Churchill.

From his military days in India, the Sudan, the latter Boer War and World War I, Churchill had amply demonstrated courage, tactical, and leadership skills, but his exemplary days were during World War II. After the war, Churchill remained in prominence, but his marked historical glory remained in the past. For the time, however, the British post war populace's attitudes of ingratitude had shown by not electing Churchill as a follow-on Prime Minister. Possibly, the post-World War II populace saw him as equal to the tirade of Hitler but too bureaucratic for rebuilding Great Britain. As any politician can attest, reading the fickle postwar population can be impossible. The British Empire remained intact through Churchill's leadership, but gratitude in politics is a fallacy. Later, when the nation needed leadership in a prime minister, Churchill was reelected for his problem solving abilities.

For several hundred years, Great Britain has had a policy of imperialism with many satellite nations or countries to list: The colonial Americas, Canada, Australia, New Zealand, South Africa, India, Singapore, Hong Kong, the Falkland Islands, and Africa are but a few of the territorial holdings of the country. During Churchill's time, he fell in line with and perpetuated imperialistic thinking. For example, he was adamantly opposed to Gandhi and his organized, relatively peaceful movement for India's independence from Great Britain's rule. According to Churchill, a grant of dominion for India would cause civil strife to both England and India. This reasoning proved fallacious, since the interrelationship of the two countries was based on rule, rather than significant trade of goods, materials, or customs.

Before Churchill's birth in 1874, British Imperialism was in full vigor. From his upbringing and schooling, Britain's empire was tantamount in his education and learning. It is not surprising, then, that Churchill believed with his heart and soul in British imperialism and that the sun should never set on the Empire. As with the Senators of Rome, the policy of an ever expanding empire has limits of rule. In Great Britain's period of expansionism, ruling egos either could not see the limits concept or chose to ignore the idea. The height of British fashion was Empire expansion for a few hundred years. Churchill was bathed in British imperialism, as demonstrated by his speeches.

Great Britain is and has always been a great country. In the couple of centuries before World War II, the Empire was beginning to feel the burdens of nearly endless territorial expansion. India and Africa, among many countries or territories, were genuine sore spots with British leaders. One wonders: why did the British see fit to occupy and rule India? During its occupation unrest, dissatisfaction, injustices, protests, and riots were all common; yet, the British attempted nullification of these widespread social problems. Such actions cost lives on British and Indian sides and untold monies for army maintenance. Expansionism and rule became insidious and overpowering problems for the British Empire that eventually led to the relinquishment and control of many territories.

In days of old, where were the British going with their expansionism, Empire policy? Was the goal world rule? Did the English rulers learn nothing from failure of the Roman and Egyptian empires? In any country, leaders cannot function correctly if learning from history is not honored.

INDIA AND GREAT BRITAIN

History reveals Great Britain has invaded and establish a military presence in 171 countries. However, a look at some of the later years of the 1858-1947 British rule of India is warranted to show some inhumane British decisions. While there existed a penchant for nomenclature in rule, the Indian Empire evolved from the British East India Company and was transferred to Queen Victoria in 1876. The Empire of India lasted until 1947—an impossibly long time for Indians to be subjugated. The transition of the Empire of India to two states was in effect difficult for the people. The northern part became the Dominion of Pakistan and the remainder became the Republic of India. Later the eastern part of the Republic of India became the Peoples Republic of Bangladesh. By 1886, upper and lower Burma were parts of the Indian Empire, but by 1937 Burma was autonomous in British rule.

The 1947 partitioning of India and the termination of British rule was nothing short of disaster for the Indian population. The Indian Empire was populated by Hindus in Pakistan and by Muslims in what is now India. Cyril Radcliffe, the British representative for the 1945 partition of the Indian Empire, hurriedly drew the border between the new India and Pakistan countries. Chaos erupted in the population for Muslims vacating to Pakistan, and a similar situation existed for Hindus trying to get to India. As each religion's followers attempted relocation, 30 million people were in dangerous transitions and half a million were killed. Just a little planning by the British could have saved confusion, animosity, and disorientation of people trying to get to their supporting country.

In the annals of high handed world-class slaughter of helpless civilians, the British Amritsar Massacre on April 13, 1919 must be near the apex of immoralities. Millions of Indians were against British orders forbidding nonviolent gatherings, speeches, strikes, or government slowdown functions. But several thousand Indians gathered at Jallianwala Gardens for cultural and religious speeches. Brigadier General Reginald Dyer had his own introverted interpretation of British orders and hierarchical sentiment about allowing large Indian assemblies. While a speech of protest was in process, General Dyer formed his troops in the garden in front of several thousand Indians. The people were given no warning, and before anyone could escape, the British troops were ordered to open fire, which they did without remorse. In ten minutes of continuous fire, 379 people were killed and more than 1000 were injured. Women and children were dutifully killed with 100 drowning in a well as they sought safety. Wanton, directed killing of civilians in a peaceful mass gathering that was ordered by a General officer of the British Army demonstrated a clear breach of humanitarian thought and a questionable British process that allowed Dyer to be a General or to use his rank.

With the news of General Dyer's directed massacre reaching England, the English public proclaimed Dyer a hero and savior of India. From where does this type of thinking or attitude evolve? The Indians were subservient to the British Empire and, at worst, had passively boycotted or participated in British company strikes or marched for their pursuit of independence. From an Indian point of view, there was certainly a question of why the British were trying to govern India. The British live on a small island about halfway around the world, and Indians are quite capable of governing their own very large country. In any event, General Dyer was acclaimed by the House of Lords but chastised by the House of Commons. Subsequently the misdirected British public raised money for Dyer and his punishment was a normal retirement.

During 1943 and World War II, a large famine in east India and Bangladesh killed nearly 4 million Indians. Largely, the blame was placed on an incompetent British governor who was preoccupied with the war. While ships with food supplies became available, they were directed to troops already supplied. Although Winston Churchill was a great British leader and Prime Minister, he was also a proponent of British imperialism and not for India's sovereignty. He felt India should remain under British rule. He certainly was not going to redirect shipping to eastern India, near Bangladesh. In fact, Winston Churchill did more than stop ready supplies. He blocked US and Canadian ships from delivering food and would not allow Indian self-help with their own ships. Condemned to fester in the indignities of famine, Churchill inflated the price of grain such that Indians could not afford to buy at their own markets. Upwards to 4 million people endured death by starvation, while Deli telegraphed Churchill of the tragedy unfolding in India. In his reply, Churchill wondered why Mohandas Gandhi hadn't died yet. Only twenty-two countries in the world have not endured the leadership and military presence of the British. For rule of this many countries, leadership may be abrupt and callous, and the starvation deaths of millions demonstrate neither leadership nor kindness to fellow humans.

Great Britain is and was a great sovereignty, and its history reaches back in time for several hundred years. In 1497, King Henry VII sent John Cabot exploring for trade routes across the Atlantic, and by the 16th century, Great Britain was interested in the colonies in what would become America. Aggressive exploration and colonization made Great Britain an empire by the late 1700's. While Egyptian construction of the Suez Canal started in 1859, Great Britain was the largest shareholder of the canal by 1875. After the invasion of the Suez as by Great Britain, France, and Israel on October 1956, Nasser closed the canal by sinking nearly fifty ships in it. By April 1957 the canal was reopened.

As politics and struggles for power vacillated, the Six Day War closed the Suez Canal on June 5, 1967 by marooning fifteen ships in Great Bitter Lake on the order of Egyptian President Nasser. Finally, after the Yon Kipper War, a joint US and British effort had the canal partially cleared by 1974, and by June 5, 1975, the canal was 99 percent cleared. Egyptian President Anwar Sadat was aboard an Egyptian destroyer, the first ship of a convoy. Interestingly, while marooned, the fifteen ships remained crewed for eight years and had their own closed society.

The Suez consternation was near the end of the British strife for a worldwide empire. Home country resources were needed for homeland survival, and the British citizenry were concerned with country while tiring of thoughts of empire. For a small country, resources were limited and their capabilities of world dominance were diminishing. The motives of British dominance and control of 177 countries must have been greatly desired by previous leadership—but an impossible feat. History has taught that such a monumental task is all but unobtainable. The Suez Canal rule and subsequent loss was understandable for British foreign trade and macro economy, but in looking for British reasons to rule India, only insatiable desires for world domination and control surface as reasons. The insatiable desire for a world empire reaches back several hundred years in British history; demonstrating the desire for world rule is not new. What fans the fire of worldly control? Since the early days British Empire was very small—two islands, Kings, Queens, and the populace may have craved a world presence larger than the home islands. The beginning British navies were world powerful and made the British demeanor identifiable in all cultures and quarters. History also reveals world rule ambitions are not sustainable or prudent, and established control will eventually break down over the oppressed people. With time British judgment and ideas of world dominance have proven themselves fallacious, while the overzealousness, haughty and superior attitudes over the conquered peoples became the demise of British rule. Centuries of British rule prevailed, but world destiny has not been determined by it.

COLD WAR

At the end of World War II, repatriation and reorganization of Germany were in dispute by the superpowers of the US and the Soviet Union. The time span 1947-1991 describes the Cold War. Great expenditures, military buildups, and disputes over sovereignty existed between the superpowers. On a world scale, this period contained many skirmishes or local wars pitting opposing capitalistic and communistic ideologies against each other for territorial rule. The amounts of effort and expenditures that the superpowers were willing to sacrifice were nearly without equal. The US goal for Germany—East or West—was the establishment of freedom and democracy. For the Soviet Union, the goal was control and the establishment of life under communist rule.

Although human rationale should have prevailed, the superpower constituents engaged in military and weapons buildups. In particular, nuclear weapons and delivery methods were emphasized to the largest degree possible. The buildups were to produce deterrents so an attack would produce mutually assured destruction with the retaliation—M.A.D. The US was at a point where communication and reason was not effective with the Soviets, since their goals were set on acquisition and spread of controlling communism backed by a nuclear delivery system. With no early stage negotiations or compromises, the US had little choice but to match or exceed Soviet armament. The arms race had begun.

The time span of the Cold War saw many meetings of leaders from sovereign nations, mostly the US and the Soviet Union, but egos, isolationism, desire for power, control, and other factors kept the irons hot on enhanced arsenals, especially nuclear. With communist dealings in the Chinese Civil War, the Korean War, the Suez Canal Crisis, the Berlin Airlift, and the Cuban Missile Crisis—producing the closest nuclear situation since Hiroshima and Nagasaki—weapons arsenals were on an ever-increasing spiral. To stretch tensions and diplomatic relations even tighter, the Vietnam War provided the best nuclear catalyst, since the end days of World War II. Fortunately, and with restraint, that scenario did not occur. For some time, France, Russia, China, Japan, and the US have had dealings in Indochina. Japan needed resources earlier and, later, communist Chinese under Russian influence wanted Vietnam for expansion and communist control.

Through the late 40's to the 60's, the superpowers had more conferences over armament and disarmament, but the results toward nuclear peace were minimal, until the SALT talks (Strategic Arms Limitations Talks) began in 1972. As evidenced by the Soviet Berlin Blockade from June 1, 1948 to May 1, 1949, which was to stop all supplies to West Berlin, the communists were willing to sacrifice any number of humans to attain their goal of control. Only the combined, concerted efforts of the Allies kept the Soviet Union from taking control.

With a nation such as the Soviet Union forcing an interdiction on food and necessary goods for life, one must wonder what the fate of the world would be under the Soviet thought process. A realization in the world of such thought would result in a demise of mankind. Yet the goal of the communist hierarchy is complete socialism for the equalized proletariat masses, except for high-ranking, controlling communist leaders—what a life to lead. Progress under a communist or socialistic regime is an oxymoron, and sovereign nations cannot go there.

Early in 1960, President Eisenhower had a golden opportunity to meet and deal with Fidel Castro on his trip to Washington. Eisenhower's reason for skipping the meeting with the young communist revolutionary leader is unclear, but it is possible some of the troubles with Cuba may have been ameliorated. A golden presidential opportunity was missed. Instead Vice President Nixon dealt with Castro. Although not official, such news of a laissez-faire attitude toward Cuba must have arrived at the Kremlin fairly soon, possibly, indicating that the Soviets had a free hand in Cuba dealings—maybe, weapon placement. As history dictated in 1962, President Kennedy was forced to deal with an aggressive Soviet attitude on their nuclear arms installation in nearby Cuba with a do or die ultimatum. Khrushchev backed away from an imminent confrontation in exchange for no invasion in Cuba. In addition to Korea, the Cuban Crisis brought the world close to a nuclear exchange. Human thinking, especially on the part of the Soviets, nearly brought world calamity—some destiny.

Within the Army Air Force, the Strategic Air Command, SAC, had its beginnings, and its focus became nuclear strategic bombing capability. As the perceived need for nuclear deterrence heightened, the B-47 was produced in 1951, and by 1953 the aircraft was operational within SAC. Aircrews were trained on long-range flying and strategic bombing. The USAF SAC goal was the capability to maintain a twenty-four hour, 365-day alert force with the B-47. Refueling tankers did likewise. As required, the US was willing to pay the price for nuclear capability and deterrence—by necessity.

To enhance nuclear deterrence, the B-52, a longer-range, eight engines, strategic bomber was developed with manufacturing and operational difficulties, but by 1955 the bomber entered the Air Force and was operational by 1956. By 1958, the bomber was flying twenty-four to twenty-six hour airborne North Atlantic "Chrome Dome" missions. The operation was long and dangerous but provided a real deterrent to a Soviet preemptive nuclear attack. However, a B-52G fire and crash over Thule Greenland signaled the end of the twenty-four-hour alert missions.   While there was some overlap of flying for the B-47 and B-52 bombers, 1955 began to see twenty-four-hour ground alert for the B-52s. The ground alert deterrence lasted for thirty-six years, until late 1991—quite a stretch.

For the span of the Cold War, the US had little choice but to meet, and preferably exceed, Soviet arms and delivery capability or the US could face possible annihilation. It was not a time for statesmen to bicker about the high price of defense to deter a nuclear holocaust. From the observation of the fanatical Soviet leaders, each was capable of providing the US a conquering, debilitating, capitalist-destroying nuclear strike, given the opportunity. Directing funds to a congressman's favorite enterprise was not how the US needed to spend money. Humans thinking in terms of communism and socialism could have brought the cold war to a hot war confrontation, if US presidents had not presented strong armament and annihilation capabilities. The nation and leading statesmen did not seek nuclear arms for war, but such deterrence measures were required for international peace and survival. The Soviet attitude has not changed; only the collapse of the infrastructure from excessive, offensive, world conquering armament expenditures brought the communist thinkers temporarily to reality. The thinking of world domination is not extinct. For now, the philosophy of dominance is unfunded in lieu of national longevity. So we must wonder where the future world is going. More communication and a deeper understanding of international relations are required for all world statesmen.

From 1959 and the beginning of the Vietnam War, the superpowers began great conventional weapon expenditures in addition to cold war expenditures. Initially, the Soviet Union had contributed well over 400 million equivalent dollars in supplied arms and over 15,000 advisors to North Vietnam. China followed suit with over 300,000 troops and something short of 200 million equivalent dollars on arms supplied to North Vietnam yearly. But over the years to 1992, the overzealous and over-budgeted control plan that included the Middle East and Afghanistan led to an internal collapse of the Soviet Union. Lack of historical leadership led to a depletion of monies and disintegration. Gorbachev was trying to hold the union together, and finally in 1990, the Communist Party was forced to relinquish its powers of control. To heighten the internal turmoil and weakness, several Soviet Republics had threatened to secede.

From the days of Lenin and Stalin to Khrushchev, Brezhnev, and Kosygin, the citizens of the Soviet Union had to endure control, expenditures, and hardships of attempted world control under the grandiose thinking of their godlike communist leaders. Even now, the Soviet or Russian leadership would return to Leninism if it was monetarily possible. With this latent, socialist-thinking bent on world control, one must wonder about the direction of the world.

Korean War

The Korean War was a conflict between North and South Korea that forced China, Russia, the US, and the United Nations to either demonstrate or display their political philosophies. North Korea relied on the sympathies or support of its communist allies. South Korean thought was more in line with the US and UN policies and politics. As the war progressed, twenty-one countries of the UN contributed something to the South Korean war effort. The US was by far the biggest contributor to the war.

From 1910 to the close of World War II, Japan ruled Korea, but with Russia declaring war on Japan in August 1945, Russian rule began north of the thirty-eighth parallel. After Japan's surrender in World War II, UN sanctioned US forces occupied South Korea. Such a split of the nation could not support a healthy relationship of the newly minted, counter philosophies countries. In what was one country, two close proximity, counter-thinking leaderships and countries came into existence. Conflict was inevitable, and since the countries had been a single entity where north and south relationships existed, a severance would further amplify atrocities.

For humanity, it seemed a higher level of political thinking was in order. But greed and the quest for power led North Korean forces—supported by China and Russia—to invade South Korea in June 1950. Many potential contingencies existed, but the political thinkers of China and Russia could see nothing but expanded communist territory and leadership. To them the mayhem and suffering created was worth communist expansion. By late June 1950, the UN Security Council had little choice but to dispatch UN forces to Korea. The US supplied nearly 90 percent of the UN forces. At this point of the Korean War, the time for brilliant thinking had passed, the UN forces and countries were stuck in a mess, which could not easily be resolved. History has certainly verified the dilemma.

The Korean War was marked by attacks, counterattacks, border limitations, logistical problems, and easily forgotten resolutions. As leadership tried to deal with communist infringements, soldiers paid the price. In August 1950, the President obtained money from Congress for the Korean War, and in a first effort, President Truman called for a shipping blockade of North Korea. However, the President did not understand that he had no warships to carry out such an action—another sterling example of a communication gap between the President and the Joint Chiefs.

How was it that the President did not know of the naval deficiency? What would have been the consequences had a foreign entity attacked the US home front? Secondly, why has it been the devoted efforts of politicians to downsize the military post-any war? Usually, they want to show a decrease in money expenditures for votes, and to accomplish this they are willing to decrease our defensive posture, sometimes discreetly and sometimes not. These popular but dangerous vote getting tactics have and will jeopardize the country.

From prior times with Korea, lack of communication among government departments was prevalent. Because of defense and material cuts not known by the President and others, such as Omar Bradley, a reorganization of initial fighting forces in Korea was necessary. At this point, even poor communication would have given a clue to the situation, rather than the worthless, existing intelligence void. Because of the governmental communication deficit, American troops were forced to fight a rearguard defensive action due to the logistics shortfall.

Through bureaucratic shortfalls and lack of communication, US troops, working through the UN, still kept an upper hand with attacks and counterattacks. A clear victory was not in sight for UN and US forces, and to correct the situation President Truman and General MacArthur had independently considered using nuclear weapons. With China, Russia and North Korea in different roles, the use of nuclear weapons on North Korea would surely have been tantamount to escalated war or nuclear war, with possibly limitless boundaries. This idea pervaded through the thoughts of the President, General MacArthur, and the Joint Chiefs. Under the existing limitations and the power of China, Russia, and the US, the use of such power was not unwarranted—but close.

The war was fought with many stalemated skirmishes, and the armistice negotiations proceeded similarly. For an unbelievable two years, talks centered on fruitless matters while people died. Initially, Kaesong was the location for negotiations and then Panmunjom became the center for talks. The communists were fighting a battle of attrition concerning troops, logistics, armament, and, possibly, morale. Soldiers were dying while peace talks ruminated about the size of negotiation tables. In reality, the communist negotiations were concerned with communist POWs held by UN forces who did not want to repatriate to North Korea but instead wanted to remain in South Korea. Communist negotiators could lose face over such a defection from a so-called "perfect" world.

From massive deaths, poor leadership decisions, indecision, the US's popular but strategically poor defensive budget cuts, governmental communication gaps, and poor perception of the world situation, what learning came from the Korean War? Before Korea and after World War II, US politicians should have learned something other than the requisites for getting elected to office. Countries and the human race pay the price for lack of foresight in leadership, which is sacrificed for popular, vote getting, no-guarantee political innuendo. In retrospect, at the end of the Korean War, we as humans or a nation have learned little or nothing and, sadly, few of us know that we don't know.

During the Korean War era, the sophistication of US intelligence gathering was little improved from World War II days. This intelligence gathering deficit, combined with an ongoing interagency communication problem, left our leaders with few tools with which to make timely, viable decisions. A retrospective view of the war and decisions, hopefully, gives us academic learning but, of course, we cannot change history—much as we would like.

Kennedy Assassination

During US history, the country has suffered assassinations of four presidents and attempts on six others. For the time, John F. Kennedy's assassination holds the most interest due to how recent it was, although more than half a century has passed. Many authors, scientists, and investigators have studied, replicated, and researched the assassination but found very little. Guilt, for good reasons, has been assigned by some authors. However, at least two constants prevail in the US population: not many people agree with the conclusions of the Warren Commission investigation, and many people believe the assassination was a well-planned coup d'état. The intent here is not to analyze evidence, which is ample, but to look into the group thinking that reasoned a presidential assassination was a solution to the country's political problems.

Findings and evidence since the assassination point toward nullification of the Warren Commission's single assassin verdict and show a conspiracy. What are the psychological references that dominate or control the thinking of otherwise rational humans to plan and execute an elected president? With scrutiny, it can be seen that the thinking is more dangerous than the assassination. What if such thinking is used by rational people to solve other national or international problems? The reinforcement for leaders and others in a conspiracy group must encompass control, power, authority, wealth, position, or advancement—either tangible or intangible. Whatever the motivation, the reward must be great, as is the power to hide past or present coup d'état actions, evidence, or records.

If a coup d'état is the real motive for the JFK assassination, then the powers for subverting truth have longevity exceeding fifty years. History now proves the power. Details and evidence of the JFK assassination have been kept secret for years, and, as one can see, there must be reasons for the locked file syndrome. Incrimination, loss of power, and loss of control are strong motives coupled to sophistication, technology, labyrinths of bureaucracy, and protective regulations have kept crucial evidence carefully sequestered. Have we advanced from the Roman days of emperors and aggrandizing senators? Not much.

With critical evidence and documentation carefully and secretly locked away, it is abundantly clear such actions were not the manifestation of a free, democratic people. Are prolonged secrets to be maintained because the perceived ignorant masses might demand reprisals and immediate, serious corrective actions? With the Kennedy assassination well past the fifty-year mark, it is inconceivable the US has neither solutions nor corrective actions for the mayhem committed. With such a format, what is our destiny?

Vietnam

In some ways Vietnam has links to World War II, since it offered rubber, minerals, and other war making materials to the supply a starved Japanese nation. The interim from World War II produced a communistic state of China which had inculcated North Vietnam with communist ideals. Of course, not all the North Vietnamese populace were communists, but survivability became a prevailing factor. Ideally, it was a case of a few powerfully situated controlling the masses. In the history of mankind, the situation has repeated itself many times—always a few want to control the many.

Exact dates are perfunctory, but communistic interest in South Vietnam manifested itself in the late 1950's and grew stronger with time. In 1949, southern Vietnam was known as the State of Vietnam, essentially causing a political bypass of the Joint Chiefs. As the Joint Chiefs became more political and less militaristic for survival on Capitol Hill, the 1954 Geneva Conference recognized the country name South Vietnam. From the time before World War II and as a colony known as Cochin China, South Vietnam was administered as part of French Indochina. However, as communism gained strength in the 1950's, interest in the acquisition of South Vietnam dramatically increased.

From the late 1950's, the US saw the spread of communism paralleling aggression and expansionism. The 1964 Gulf of Tonkin incident in which a US destroyer clashed with a North Vietnamese fast attack craft cemented the US involvement in Vietnam. Leadership in Washington had several options for dealing with foreign country aggression, but 1965 saw US deployment of troops and the generation of a multi-border air war. 1968 saw a peak in warring activities with fighting continuing until August of 1973. Depending upon a chosen start date, the war lasted fifteen years, maybe more.

From a participant's point of view, politics and politicians had more to do with prolonging the war than ending it. Generals and admirals were in complete obfuscation. From tactical decisions made in Washington, military leaders were but pawns in the war. Their expertise did not appear utilized. Instead, high-level politicians with no military experience began picking tactical air and ground targets with VC body count used as markers of success or failure. Apparently, there was no goal set for determination of overall victory or failure. If a goal or success is not set, how is victory or defeat determined? Politicians look for immediate success or victory in order to shine for constituents and contemporaries. Re-election and reappointments are forever in the political forefront. Unfortunately, throughout the Vietnam War, generals and admirals took a backseat to the "all-knowing" Washington political minds. Therefore, the war lasted many years beyond a possible true military culmination. Has this type of political thinking changed? Sadly—no.

The term "politically correct" had been around for some time, but in the Vietnam War, the term was politically operational. For the most part, generals and admirals disdained the term. To military leaders the term morphed into rules of engagement or ROEs. Under ROEs, certain valuable tactical targets were off-limits to ground or air warfare. Allowing targets of materials, munitions, guns, explosives, or other weapons to survive in cultural or religious areas, was a certainty to prolong the war and send a message of "we'll keep hands off" to the enemy. These political sanctions contained in the form of ROEs were difficult, if not impossible, to overcome from a tactical viewpoint. Military leaders had their hands tied. If a war is to be fought, then to save lives, military leaders must operate under military rules and decorum, not political rhetoric and inappropriate, incorrect political goals.

From the early days of the Vietnam war, it was abundantly clear that the bombing and closure of Haiphong Harbor would cripple supplies to the North Vietnamese regulars and Viet Cong to such a degree that Ho Chi Minh would capitulate. The major supply port would cease to exist, while rusting hulks in the harbor entrance would act as a long-term deterrence. ROEs and the Washington idea of "politically correct" kept the Air Force from bombing Haiphong

Harbor early in the war. Such an action would have saved many lives and a very lengthy, expensive war.

To the military's disadvantage, the administration considered the nationality of some ships in the harbor. They were not all Russian; a "politically correct" attitude was in force, saving enemy lives. As the air war escalated in Vietnam, bordering countries such as North Vietnam, South Vietnam, Cambodia, and Laos were sovereignties of concern—were they ever. Along the Ho Chi Minh complex of trails from China to these countries were interdiction points. Some of these points were crucial while others were regular, mundane targets of questionable tactical worth. Reconnaissance and intelligence of targets were filtered by Blue Chip or 7th Air Force, but Washington had a hand in selecting interdiction targets as well. Scheduling a lucrative target or interdiction points through a labyrinth of Washington's and 7th Air Force's approval channels could render the target worthless just through the passage of precious time. As the Commander in Chief had indicated—the targeting of an outhouse would not be scheduled unless personally approved. Operating under such an attitude of leadership severely handicapped the military for victory, maybe preventing victory altogether. It was understood that the Commander in Chief is the head of all military forces, but this position does not imply expertise in military tactics. Since the President is a politician, the government appoints the Joint Chiefs—all generals and admirals with the expertise so desperately needed. Since military actions and operations needed to be "politically correct", this essentially caused a functional bypass of the Joint Chiefs, forcing the Chiefs to become more political and less militaristic for survival on Capitol Hill. For future national survival, this country must have strong representation in the Joint Chiefs and the Commander in Chief must listen to their advice.

The demilitarization plan signaled the de-escalation of the war for the US and increased Vietnamization, which would not be strong enough to repel North Vietnamese and Chinese Regular forces. In 1973, the release of POWs began under total communist control and scheduling—not US. From the middle 60's until 1973, our negotiations on prisoner release had been, at best, pathetic. At the negotiation table we lacked prowess and power, as indicated by the seven or eight years some of our POWs had to languish in Hanoi. These years were not easy, as they became filled with torture, interrogations, deprivation, starvation, and atrocities forbidden by the Geneva Convention, which the North Vietnamese completely disregarded. How does a nation's credibility stand by letting communist thugs abuse US soldiers as prisoners for eight years? How does one explain our lack of forcefulness when dealing with the North Vietnamese? Was "politically correctness" a factor—it was.

By 1975, complete evacuation of US forces and personnel was necessary. We could not leave fast enough, and we were forced to depart without many faithful Vietnamese personnel. There are many facets from which we should have learned, compared to the little knowledge we gained from fifteen years in Vietnam. Should we have had better intelligence on the Indochina/Vietnam situation? We absolutely should have procured better intelligence from our agencies and reconnaissance. We had little idea of the outstanding entrenchment the Communists had in the hills, caves, and on the Ho Chi Minh trail. As the US became more involved in guerrilla warfare, millions of dollars and sophisticated weapons were being used to produce a kill or probable kill on a few VC hiding in jungle caves. The communists were winning the war on attrition of weapons, soldiers, and funds—worse yet, the US did not realize it.

From the point of view of top US ranking politicians, egos and popularity must be divorced from good, workable war strategies. The Joint Chiefs must contribute, be heard, and recognized as the resident experts on warfare; their communications must be directly to the Commander in Chief—not bureaucratically layered to some underling politician. Politicians must realize the Joint Chiefs have at least a hundred years of combined military experience. Since Congress approves promotions to general and admiral, care must be taken by Congress that such promotions are based upon military competence, not political popularity or connections—a difficult task requiring time for consideration. From the Vietnam era, if not other wars, interservice communication was difficult, because service rivalries had been predominant. The Joint Chiefs must be tasked with ensuring interservice communication, cooperation, and that all services espouse and engage in team work.

Presently, and from this time forward, it is not likely the Commander in Chief will have any military experience, and for this reason there should be clear, direct lines of communication from the Joint Chiefs, President, and Secretary of Defense. War records of the past indicate loss of life due to decisions based on "politically correct" derived ROEs, or rules of engagement. In war, following politically derived ROEs can prolong war and cost lives. If ROEs are prevalent enough, maybe, the US should either rescind them or not be engaged in warfare. To ensure the US does not get bogged down in skirmishes or war laced with ROEs, the President and Secretary of Defense should listen to and heed the Joint Chiefs. In this inner circle of confidence and exchange, politics should play no part—a difficult parameter that only true leaders can follow.

In retrospect with the Vietnam War, the nation needs to review the attitude and disdain many US citizens had toward their soldiers. Many citizens were not happy with the government's involvement in the Vietnam War and transferred their disdain and hatred to returning soldiers of the time. Such a transference of hate does not show a lot for rational thought. Many civilians don't realize that military personnel are arms of the government and act according to their commanders and the Commander in Chief. They are under orders and do what they are told. Soldiers do not play in the world political whims, and in the Vietnam era, soldiers were acting on two centuries of military precedence. But in the eyes of many civilians, soldiers became the focal point of hate and frustration with the government. Many citizens in the 60's acted with ignorant emotion rather than intellect toward fellow Americans.

In the term "soldier", Marine, Navy, Army, and Air Force personnel are all included. Many US citizens wanted to place psychological transference of wrongdoing or other government maladies on their soldiers. No other term but ignorance applies. All soldiers are proud of their accomplishments and service records, but in the shallow thinking of the 60's, military men and women would substitute civilian clothes in a civilian environment to avoid, if possible, dispersions of some misguided civilians. For the civilian hatred displayed towards the military in the Vietnam era, this nation has shame to bear. Those civilians, the few supporting the military, deserve credit. Those not supporting the military deserve disdain and discredit, celebrities or not.

As we should know, a sovereign nation needs a strong military contingent—all branches. For modern times, the military is to deter aggression, rather than be for offensive purposes. How does a formidable portion of the civilian populace grow counter to the protective military? Many of the college age people in the 60's had little or no sense of obligation. Although they had rights, freedoms, and privileges granted by the US and its Constitution, they felt entitled with no obligations or responsibilities. With a little introspection, some college age students have the same thoughts now. To remind the thoughtless, soldiers gave citizens and students the freedoms and rights—not Presidents or statesmen. From the Vietnam era, the US has a myriad of draft dodgers ducking a real service obligation. These people felt above such mundane service and saw no sense of responsibility. Yet, after the Vietnam era, most of the draft dodgers wanted full citizenship with the associated freedoms and privileges, which were generally and wrongfully granted. One must ask then: in any hierarchy, where does such a forgiveness policy place those who served when called? What emphasis will be placed on responsibility and loyalty when future skirmishes or wars arise? What kind of citizenship are we teaching our younger generations? The answer seems to be that the qualities of discipline, loyalty, responsibility, honesty, and dependability may be returned to the country in lesser degrees than desirable. Strong sovereign countries must teach, inculcate, and educate their young citizens to subscribe to these values. If the younger generation does not possess these values, the nation will have serious sovereignty problems. What is our direction now? What values or lack of values will determine the fate of the nation?

Changes come slowly, and there is no longer a draft. However, the country has kept the right to call a draft for soldiers, and if a draft were invoked, the response rate would probably be surprisingly low. The old adage "freedom is not free" might be invoked more in our secondary educational system. The younger generations need to more fully understand the freedoms they enjoy were hard won and not simply granted by statesmen, politicians, or even Presidents. The Constitution defines and codifies our rights, freedoms, and privileges, but first, these qualities of life must be won and, then, steadfastly maintained. In our secondary educational system, a learning from history as to citizens' rights and responsibilities might give the nation some direction as to where we are going, while producing sovereignty through good leadership in times to come.

Pueblo Incident

On January 23, 1968, North Koreans captured the "USS Pueblo" with eighty-three crew members, causing an international high seas incident. At the same time, the US was deeply involved in the Vietnam War. While the US Navy maintained the "Pueblo" position in international waters, the North Koreans stated otherwise, and by international agreement, international waters begin twelve nautical miles from country shores. North Korea claims a fifty nautical mile sea border on its own authority, which is not recognized by other countries. The capture process cost the lives of two sailors by North Korean guns. In eleven months, the crew was repatriated to South Korea after torture and endless interrogations of the crew. The "USS Pueblo" still remains under North Korean authority moored on the Botong River in Pyongyang. Is leaving a US naval vessel under the control of North Korea the actions of a super-power or a sovereign nation? Definitely not.

Although the "Pueblo" had maintained radio contact with Naval Security in Kamiseya, Japan, and the 7th Fleet was aware of the situation, rescue or air cover for the "Pueblo" never materialized. It would seem a sovereign, mighty nation at war in Vietnam with war materials at hand would have a retaliatory mission generated within, at most, a few hours. But no, the White House consensus was to keep revenge reaction to a minimum for fear the North Koreans would kill the "Pueblo" prisoners. Our leaders wanted more information so the US would not make a hasty response. Soldiers, marines, sailors, and airmen are always at some risk of dying, which any sovereign nation wants to minimize.

Although our political leaders and Congress tended to slough off any loss of credibility or respect, the US lost face and international credibility in lack of action. The basic question: how does a mighty, supreme power allow a third world communist dictator to dominate decision-making? Major General Pak, the chief North Korean negotiator wanted an admission to spying, an apology, and assurance that spying would cease. General Woodward, the chief US negotiator, found these confessional criteria foreboding to the thinking of a sovereign nation. The "Pueblo" was in international waters, not North Korean waters as claimed by General Pak. Essentially, General Woodward had to confess to actions not accomplished by the "Pueblo" for the return of sailors. False confessions by any military man are contrary to the values of leadership. Of course, the North Korean objective was to shame and demean the mighty US while extending the negotiating process of North Korea. The lives of the "Pueblo" crew and destiny of the ship were caught in the middle of the bickering emanating from General Pak.

Clearly, the situation would not have existed if the US had acted appropriately during the Korean War. When a nation gets involved in a struggle over national ideals and communist principles surface, the communist, tail-wagging, world-aspiring, dictator country must be put down. During the Korean War, the US failed to do this. Then, as well as the "Pueblo" incident, the US failed in negotiations and asserting the rules and rights of a freedom loving country. Because of indecisiveness during the Korean War, North Korea has been a thorn in the side of civilized society since 1953. The monies spent by the US to guard South Korea have been without a finite end and far more expensive than a successful, victorious termination of the Korean War. An established freedom loving form of government would have provided one less country for communism to propagate.

As during the Korean War, the "Pueblo" incident was riddled with politically correct, tainted thinking. Leadership did not want to arouse a homeland population bent on retaliation, nor did US leadership want to further antagonize North Korean leaders for fear of reprisals on prisoners. It is difficult not to oversimplify the situation, however, the capture of the "Pueblo" and the imprisonment of its crew is not complicated. North Korea was a country wanting power and international recognition. The solution to North Korean aggression and zestfulness for power was either to divert a couple of B-52 sorties to the port of Wonsan for an accurate bombing of the "Pueblo," since no people but North Koreans were aboard. More desirable would have been the placement of a US battleship in a protected position for sixteen-inch shelling of Wonsan. Cruisers and destroyers could have participated in the shelling and protection of the battleship. Such treatment for haughty, arrogant, power-hungry leaders of North Korea would demonstrate a promise of future retaliations and bring the country into something closer to civilized behavior. Fortunately, for the North Koreans, "politically correct" was a big catalyst for letting the North Koreans get away with high seas hijacking of a US naval vessel and murder.

So much for proposed cures of the "Pueblo" incident. What remains are two items of shameful evidence. First, the North Koreans kept a US naval crew in captivity for eleven months to endure numerous communist interrogations and indoctrinations. The amount of time in captivity was lengthy enough for the rest of the world to notice and make shameful observations about US softness or how no retaliatory action against a war provoking action. Secondly, one wonders how a first rate, world recognized sovereign nation could allow a sub-third world North Korea to detain a US naval registered ship from 1968 until the present. Since 1968, many US government administrations have come and gone; yet, not one has repatriated the "Pueblo" to the US Navy. This continuing situation is shameful, but we choose to ignore it. With leadership values that either condones or ignores the arrogance of the North Koreans, the world and, especially, the US is guaranteed more of the same anti-civilized, cultural contrary mores behaviors. From 1968 to present, the world has seen nothing but arrogant, self-indulgent, witless leadership from North Korea. Until the North Koreans agree to meaningful, enforceable treaties, sign peace loving instruments, and join civilization, the world will continue having problems with North Korea.

E C-121 Downing Incident

How do citizens, leadership, and sovereign countries explain the downing of an unarmed reconnaissance naval aircraft over the Sea of Japan? Justification, legally or logically, is nonexistent for the EC-121 take down and murder of thirty-one military personnel. The North Koreans launched two Mig-21 Fighters from near Wonsan. The commander of the mission, Lieutenant Commander Overstreet had a standing order for flight no closer to the North Korea coast than 50 nautical miles. Friendly radars had tracked the EC-121 until the Mig-21 fighters merged with the radar target. Within a few minutes of the shoot down, President Nixon was advised and completely surprised, since these missions had a peaceful history and warranted no reprisal actions from North Korea.

Following the attack, several responses were proposed, including an attack of retaliation initiated by Representative Mendel Rivers. The National Security Council and the Joint Chiefs had planned several physical reprisals against North Korea. Any of a dozen reprisal options would have devastated efforts and curtailed interior functionality of North Korea. In the end, "politically correct" produced a do nothing attitude. US leadership indicated no information about the location of either US or North Korea forces. So, no action was taken. We might have had a show of naval and air forces, but we resumed EC-121 missions. The rhetoric following the incident included Henry Kissinger saying, "Our conduct in the EC-121 crisis was weak, indecisive, and disorganized." Additionally, the President prophesied the North Koreans would never get away with it—again. A greater input of fear cannot be imagined.

Because the US claimed little information and difficulty communicating with the Pacific Fleet, the North Koreans got away with murder and mayhem on the high seas, but the North Koreans received the "hurtful" words from the White House. In the annals of history, the EC-121 incident will be probably noted as not a historically significant event. Yet, an arrogant, criminally prone, third world, self-professed country, successfully and without repercussions, assaulted a sovereign world power and got away with it. We didn't have enough information to counter attack; yet, the North Koreans attacked a helpless reconnaissance aircraft with little or no apparent reason. At what measure of shame and disrespect does the nation take action? The US was simultaneously engaged with guerrilla forces in Vietnam, but such an engagement was not an excuse for inaction against the North Koreans. In fact, the US had forces in the Pacific and South East Asia to discipline the North Koreans quite handily. We knew two Mig-21s had shot down the EC-121, but the information void, coupled to the usual communication problems, were parameters enough for "politically correct" inaction.

If a philosopher was to draw conclusions from the "Pueblo," and EC-121 incidents, it would not be difficult to determine positive results of the Vietnam War. The domineering principles of "politically correct" and the derivatives of ROEs, rules of engagement, handicap a nation to such a degree that victory is a waning aspiration—or unattainable. The battle is lost from within, not exterior devastating forces. An opinion consensus would deem harsher consequences for North Korea in both the "Pueblo" and EC-121 incidents. A sovereign nation such as America cannot have its vessels captured or shot down and military crewmembers killed or held hostage by a country which is flagrant and nonobservant of international rules and laws. History has amply demonstrated a continuum of country rulers who will flagrantly disregard international protocol for the sake of power, prestige, or wealth. This type of contrary cultural, self-indulgent behavior has been a given over history and will continue into the future unless methodically corrected.

A singular prescribed rule or behavior is, probably, not the answer. However, for elected leaders, the electorate should look for individuals demonstrating talents for international relations which emphasize a sovereign, strong America and propose relationships and agreements that are fair to all countries involved. Such words are easy while the actions are difficult. For those appointees and elected officials, a gratitude to supporting organizations should exist, but the onerous part requires these people to now think in terms of sovereignty and unity for their country and not the puny, frail, or selfish wants of the cronyism that helped them to office. Everyone involved—appointees, those elected, and political support organizations—must realize the national and international priorities.

Iranian Hostage Crisis

The Iranian hostage situation was a clear case of Muslim Iranian students trying to make a political point, which got out of control. They were supporting the Iranian Revolution and erroneously determined holding some US Embassy hostages for a few days would make the world see and understand their situation. Iranian consensus saw the hostage situation as revenge against the US for attempts to thwart the Iranian Revolution and its support of the overthrown Shah, Mohammed Reza Pahlavi. It was perceived that the US sanctuary for the Shah was an endorsement of atrocities he had committed. Endorsement was certainly not the US intention; previously, the Shah had been the singular person with whom the US could negotiate or work.

From the US point of view, the hostage taking was viewed as a breach of international law in which diplomatic immunity and freedom from arrest was completely disregarded. The situation stagnated until Iraq invaded Iran, and earlier in 1980, the Shah died of cancer in Egypt. Ayatollah Khomeini, who was strengthened by the Shah's departure, relished his power with the hostage control, but the Iraqi invasion forced some negotiation with the US using Algeria as the mediator. From lack of leadership, the US was forced to table talks in Algiers with the fate of the hostages in question and jeopardy. Not until January of 1981 were the hostages released, and at this point they were flown to Algiers and thence to Wiesbaden. For 444 days, the US had no control and demonstrated little to no leadership. The best the US could do was tie yellow ribbons around trees.

113

From the first embassy takeover attempt in February of 1979 and the actual takeover in November of 1979, the leadership and military had time to plan a recovery of the hostages. However, with the excuses from Washington and the Pentagon, the result of the puny rescue attempt was failure and the loss of eight US servicemen—shameful. Diplomatic and political rhetoric have found a little polish in excuses and, in retrospect, professionalism and leadership were nil and nonexistent. The rescue operation appeared as a shot from the hip, yet there was time for a workable plan. Also, the US had standing, workable forces with generals and admirals providing the necessary leadership. As history records, the rescue was a disaster, and later, a special operation review group was to study the causes. However, from a historical point of view, lack of singular leadership, lack of intelligence, and lack of inter-agency communication were primary to failure. Weather, of course, was a factor, but forecasts and transmissions to using units was vital but avoided for a radio silence condition. Our soldiers are excellent, but we were working against ourselves. It is difficult not to notice; our litany of military and political faux pas is fairly lengthy.

After the failed rescue attempt, Ayatollah Khomeini's persona was bolstered while President Carter suffered in leadership, popularity, and presidential prospects. He accepted responsibility for the failure, but some generals and admirals are due some military and political heat. At what point do we, as a nation, begin to learn and take a strong, impenetrable international stance? Are we to remain ignorant of history and the lessons therein? Apparently. The Korean War, Vietnam War, and the following incidents with North Korea would weigh heavy in producing learning situations in our international protocol. With weak, indecisive responses or no response at all to atrocities committed on American soil or on the international scene, the nation is inviting more trouble than it is trying to avoid. Terrorists, bullies, and third world empire seekers take advantage of a "politically correct" attitude but do understand a formidable defensive force directed by good leadership not influenced by political rhetoric.

Beirut Bombing

The loss of 241 soldiers in the Beirut barracks bombing by the Islamic Jihad Organization should have been prevented by experience and learning in previous wars, especially Vietnam. But between the Marine barracks at the Beirut International Airport and the French soldiers at the Drakkar building, a total of 307 soldiers, sailors, and civilians died. Has the US not learned the bitter lessons from past? We have definitely not, and the price of not learning due to the political correctness concept has been monumental when considering the loss of lives. Through political correctness and the corollary of ROEs (Rules of Engagement) our Marine sentries were not allowed to have magazines in their rifles or a chambered round. With such a concept, why do we have sentries? As the tragedy unfolded on October 23, 1983, Iranian Isruail Ascari detonated his 21,000 pounds of TNT after driving his truck through the perimeter fence, the guard shack, and into the barracks lobby. Unfortunately, the sentries at the gate were operating under the helpless rendering of the ROEs—no magazine or round chambered.

Later, under President Reagan's direction, a fact-finding committee recommended fewer lives might have been lost if the guards carried loaded, ready to use weapons instead of the useless, politically correct, not ready, safe weapons—what a brilliant, fact-finding committee finding. Although the military was a protective tool, unknowing politicians were necessary to the perpetrator. Is it not obvious, if a soldier is deemed necessary, that they should be fully armed and ready for any contingency? The leadership of a sovereign nation does not think in terms of political correctness but arms itself against the enemies of freedom. Where or when did we lose sight of the ideas and principles of a sovereign, freedom loving nation? Why are we using those in the military as pawns rather than soldiers equipped with brains? Few things are more formidable than a well-trained soldier capable of thought, but from any point of view, especially political, these precious military assets cannot be wasted.

In June of 1982, Israel invaded Lebanon for a buffer zone between the Palestine Liberation Organization (PLO) guerillas and Syrian forces. Later a multi-national force (MNF) of French, Italian, and US Marines were deployed to Beirut as a peacekeeping force. The US agreed to support the Israeli operation with arms and materials. Furthermore, the US supported the Lebanese President Bachir Gemayel and the Lebanese army. Such alliances alienated many, including Lebanese Muslim and Druze communities. The Phalangist militia were, unfortunately, linked to the US backed President Gemayel, and they were responsible for attacks on Lebanese Muslim and Druze people. To add animosity, the Phalangist militia had attacked PLO refugee camps. For the associations, the US and its military representatives were looking like targets for revenge. It appears the lack of international intelligence and plain espionage were setting the Marines up for targeted revenge. The expertise of generals and admirals was not utilized, nor the learning from recent Middle East happenings or lessons from more distant history.

Airline Shooting Incidents

From a civilized world point of view, archaic and primitive thoughts occasionally pervade the radicals of society whereby a civilian, unarmed airliner is shot down for perceived or real transgressions against the country. How does a society of modern times manage to produce individuals who can justify killing civilians by downing their airliner, no matter its origin of flight? In the minds of radicals, their cause or injustice gives reason for actions against any society. From the 1930's to present, radicals or civilized societies have committed atrocities or incidents with civilian airliners with anti-aircraft artillery or surface to air missiles (SAM) in excess of thirty times. For radicals to express their disgruntlement, innocent airliners offer availability and ease to partial destruction or complete annihilation for perceived injustices to radicals in some societies. In general, the societal extremists come under such names as "The Peoples Front", "Government Aligned Extremists", or "The Liberation Rebels." These people form a group with a leader, who understands group psychology, and contemplates a plan to express their suffering from injustice. Oftentimes the target is a defenseless airliner, which will get the attention of the news media.

The Russians have a record of airliner shootings with no real justification. On September 1, 1983 the Soviets shot down Korean Airlines Flight 007 with a Sukhoi Su-15 interceptor near Moneron Island. With little doubt, the Boeing 747 had errored in navigation, because the crew was not situationally aware and continued erroneously on a course deviation. Although all the navigation waypoints had been set in the multiple inertial navigation systems, apparently the autopilot remained in the heading mode, and the aircraft never got within the distance parameters to allow autopilot acceptance of INS or inertial navigation inputs.

The process the Soviets used for justifying the shoot down of KAL 007 was, to say at best, introverted, convoluted, misdirected, and short on international diplomacy. Historically the Soviets had demonstrated a predisposition and overprotective attitude for infringement on their borders. The Soviet political structure tends to reward aggressive behavior on border intrusions, and the rise of military stature as well as political advancement are rewards for protective behaviors—however unjustified.

From the records, four Soviet fighters were launched on May 23 from Smirnykh Air Base and three SU-15 fighters from Dolinsk-Solkol. Initially, the Mig-23 fired more than 200 rounds for a warning, but, apparently, the Boeing 747 crew did not see them. The rounds were armor piercing, not incendiary, and nearly impossible to see.

Subsequently, for fuel savings, KAL 007 contacted Tokyo Area Control Center for a climb, and, strange as it may seem, the lead SU-15 pilot interpreted the KAL 007 climb and slow down as an evasive maneuver. How is it that a well-trained Soviet fighter pilot can't understand the difference between a Boeing 747 climbing and an evasive maneuver? In reality, a huge aircraft like the Boeing 747 is pretty limited in its evasive maneuver repertoire—it is an airliner designed for efficient carry of large passenger loads over great distances.

Under political and military pressure, General Kornukov ordered Soviet ground controllers to issue the shooting. Major Genadi Osipovich, the lead SU-15 pilot, had reported blinking lights and two rows of windows, and although he knew the aircraft was a Boeing 747, he did not report that fact. With no regard to the loss of humanity on a commercial airliner, which had inadvertently transgressed Soviet airspace, Major Osipovich attained a missile firing position and executed his order from General Kornukov. The K-8 missiles were successful and hit the airliner, but it did not explode in flight. Instead, with little control, the Boeing spiraled over Moneron Island before crashing in the ocean. With supplemental oxygen, there is a high probability the passengers were alive until impact. All 269 people aboard were killed with only body parts found later. After infringing upon Soviet airspace and continued flight, KAL 007, in high probability, was actually fired upon in international airspace.

From societal mores and the supposition of a civilized country, it is difficult to find the Soviet justification for intentionally shooting down a passenger airliner, which because of navigational errors had transgressed on a remote part of Soviet territory. The arrogance, haughtiness, and disregard for fellow human souls can only be described as barbaric behavior. Soviet political history has evolved to reinforce thoughts of territorial aggressiveness. Political and military actions over time reflect an overzealous attitude of protectionism and border defense. Restrictions of Soviet citizens passing the borders with difficulty are symptomatic of travel restraints.

The Russian shooting of KAL 007 is not a singular incident. Early on, the tally of Soviet shootings included a Finland operated civilian passenger JU52-3/MGE Junkers aircraft during 1940 and World War II. The Soviets or satellite countries have continued with other civilian airliner aggressiveness. In 1978, SU-15 fighters shot down KAL 902, a Boeing 707, near Murmamsk, Russia after it had ventured into Russian airspace. The airliner was forced to land on a frozen lake. In 1993, three Trans Air Georgia aircraft were either shot down or grounded by gunfire or missiles in Sukhumi, Abkhazia, Georgia, a country heavily influenced by the Soviets. A surface to air missile brought down Siberian Airline Flight 1812, and a Tupolev TU-154 aircraft was fired upon during a Ukrainian military drill. Seventy-eight people died for the Ukrainian exercise. Near Donetsk, Ukraine Malaysia Airlines Flight 17, a Boeing 777 was downed by a surface to air missile. 298 people died for a questionable incursion into Ukrainian airspace. Regardless of country origin or navigational errors, Soviet influence on aircraft shootings has been heavy and without due cause.

Where is the world and civilization going with a country filled with this type of non-humanitarian, contrary, isolationist thinking? As the world progresses, introverted, self-centered thinking is extremely detrimental. Macroeconomics, international trade, communication, credit transfers, and business travel are some of the ingredients for world advancement toward peace and harmony. But with barbarian thinking and shootings of international commercial carriers, humanitarian and civil progress is stifled in many essential fields. The shooting down of a civilian air carrier, whether it erred in navigation or transgressed the sanctity of a border, is not justified in any culture's set of laws or rules.

At some point in the world's near future, sovereign nations need to realize the need for better communication and the utilization of only top internationally educated statesmen. The days of seeking world dominance are over. If not for humanitarian reasons that nations can propose international civility, then, the earth cannot stand nearsighted with introverted nations declaring international wars. As the human earth dwelling species, we have not yet realized or learned the art and science of international protocol and peace. History has shown empires and countries counted their success and prowess on the annihilation of the enemy forces. Targeting airliners also targets citizens from all countries. As the supposedly intelligent species on earth, we have learned little but require much. When will we learn not to jeopardize the civilizations within which we must live?

GULF WAR

The Gulf War encompassed Operation Desert Shield and the defense of Saudi Arabia, Operation Desert Storm. The war spanned from August 2, 1990 to February 28, 1991, and in that short time many operations took place. As with many problems in the Middle East, Saddam Hussein, President of Iraq was basic to the problem. Iraq had been in debt from the Iran-Iraq war with a great deal of money owed to Saudi Arabia and Kuwait. From the days of the Ottoman Empire's province of Basra, the ruling family of al-Sabah negotiated a protectorate agreement with the United Kingdom. Essentially, Iraq was landlocked and claimed Kuwait as part of and belonging to Iraq. To add frustration to Iraq and Saddam, Kuwait was accused of slant drilling into Iraq's oil productive Rumaili oil field. Iraq was further deprived of some of its revenue sources. The continued overproduction of OPEC oil by the United Arab Emirates, dropping oil to a temporary price of ten dollars per barrel, added to the impoverishment and inability for Iraq to repay. Returning expatriates who had departed for Egypt were causing problems for Iraqi citizens—probably a question of returning expatriate's Iraqi loyalty.

Saddam complained to the US ambassador for Iraq, April Glaspie, that the US was not seeing Iraq as a friend. Glaspie indicated the US did not want to interfere in Arab affairs, except that there was notice of massive troops in southern Iraq. If the UAE and Kuwait take military actions, then Iraq and the US must become concerned. Later, in the Jeddah talks, Saddam wanted ten billion dollars from Kuwait for lost revenues at the Rumaili oil field. Kuwaiti leadership offered one billion dollars short of the demand.

Since the Kuwaiti response was short of Saddam's demand, he attacked Kuwait City on August 2, 1990. For the Iraq-Iran War, Iraq had amassed a sizable army with world-class numbers in troops, boats, tanks, and aircraft—jets and helicopters. Iraq's main thrust to Kuwait was a combination of commandos dispersed by boats and helicopters, and within half a day most of Kuwait City was under Saddam's control. Subsequent to Saddam's Kuwait City take over, much political activity ensued. The United Nations Security Council immediately passed Resolution 660, which condemned Saddam's invasion and "demanded" a withdrawal of Iraqi forces. It's hard to imagine the lack of fear the resolution instilled in Saddam, considering his disdain for the UN Security Council. For Iraq, economic sanctions and trade blockades soon followed which, when considering the structure of Iraqi debt, was effective. By August 12th, 1990, Saddam began to squirm by proposing occupation charges against him be resolved. As a further condition, Saddam wanted Israel to withdraw from Palestine, Syria, and Lebanon, and to favor his position, he wanted US forces removed from Saudi Arabia with replacement of Arab forces. As an Arab idealist, Saddam wanted sanctions and embargoes lifted, no threat from Saudi Arabia, oil money for Rumaili, debt to Saudi Arabia forgiven, oil agreements from the US, and Israeli troops out of Palestine. Saddam was placing himself in and untenable, non-negotiable position.

Regardless of bargaining, the US held fast on no concessions until Iraq had completely withdrawn from Kuwait. Quite noticeable to the US and the UN Security Council was the close proximity of many Iraqi forces to Saudi Arabia and the extensive oil fields therein. Saddam was a radical thinker and terrorist, and if he was allowed a free hand at the Saudi oil fields, Iraq would have control of a major supplier of the world's oil supply. Obviously, the world does not need a dictator for an oil supplier. As a possibility for further advancement to Saudi Arabia was a debt of 26 billion dollars owed to Saudi Arabia by Saddam. The Iraqi leader wanted the debt forgiven for fighting Iran, and as further confirmation of Saddam's radical, expansionist thinking, while antagonizing the UN Security Council and the US, he declared Kuwait to be the 19th province of Iraq with cousin Ali Hassan al-Majid as a military governor. As can be seen, Saddam was a military dictator with few limits on the size of his empire. Early on, some contemplation might have been given to his timely capture or demise—not to happen.

Considering the potential Iraqi advancement on Saudi Arabia and oil fears, the US launched Operation Desert Shield on August 7, 1990. Many nations in the world were concerned about the Kuwaiti situation and Saddam's potential for major oil supply control. To counter the avarice and terrorist thinking of Saddam, the US Navy moved the carriers "USS Dwight D Eisenhower" and the "USS Independence" to the Persian Gulf for readiness on August 8, 1990. And for some heavy hitting, the battleships "USS Missouri" and "USS Wisconsin" were positioned in the area. Many combat aircraft joined the operation and most would operate from airfields in Saudi Arabia. Eventually over 540,000 troops were in position. At this point, one must consider the magnitude and expense of Operation Desert Shield to counter the radical thinking of Saddam.

Saddam's record to humanity speaks volumes of despair. His abuses to human rights were lengthy and repetitive, which included using biological weapons against Iranian troops. Worsening matters against humanity, Saddam used biological weapons against his own country's Kurdish population. For some time, Saddam had a nuclear weapons program, although Iraq lacked an abundance of raw materials or a delivery system. Hopefully, for the future of humanity, neither Iraq nor Iran will ever have nuclear capabilities. Assistance to either country in terms of anything nuclear would be treacherous to mankind's longevity.

Although US leaders stated their goal was not the capture and death of Saddam, coalition leaders chose for Saddam to remain in power, instead of executing him and proceeding to Baghdad for a government takeover. This simple action very easily could have saved the Iraqi war in 2003 and a tremendous loss of lives. Minor to the Iraqi War was the dumping of 400 million gallons of oil into the Persian Gulf, and the firing of 700 wells. What an Iraqi disregard for the environment and world.

The Iraqi war began on March 20, 2003 and consisted of a coalition led by the US. As outlined by General Franks, the objectives were: end Saddam's dictatorship, eliminate weapons of mass destruction (WMD), rid Iraq of terrorists, gather intelligence on terrorists, provide humanitarian support, secure oilfields for revenue to Iraq, and help Iraqi people seek self-government. Unfortunately, the war proceeded from 2003 to 2011 with no end to mayhem, deaths, embittered country alliances, Iraqi job losses, and, sadly, the wounding and killing of many soldiers—mostly US. In terms of expenses for the Iraqi

War, the US total was estimated at 1.9 trillion dollars with the daily cost of 720 million dollars. The wounded, the lives lost, and the monumental expense must burden tacticians, present, and future world leaders to question the worth of the war. In the end, a representative type government was to be formed, but the Middle East and Iraq have yet to experience such a form of government.

Could the world have been spared the damage, stretched alliances, deaths, wounds, and massive expenses if Saddam's capture or death had been the prime goal of the Gulf War in 1991? US leaders clearly stated such actions were not objectives. Was enough emphasis placed on quality intelligence from which leaders can make real time, future affecting, and the best informed decisions? There are no dispersions to cast on the people in the intelligence or spying business, but for many years it has been probable that budgets for the CIA and other intelligence agencies have been cut to a point where the functions operate in handicapped modes. Commensurate with modern times, any country—especially the US—cannot be without the best intelligence agencies.

For the Gulf and Iraqi Wars, a precedence and ongoing intelligence network was necessary. As reliable as possible information is necessary for forecasting adversarial actions but, in retrospect, our leaders did not appear to be armed with real-time intelligence. The US federal government is continually getting larger, and some leaders may see the need for agency or inter-agency spying. A corollary to this idea is intelligence work on the citizenry, which to the smallest degree might be necessary. Consistent subversives in the population may need some attention. However, the large budget of monies for intelligence would be more effectively spent on international affairs, especially the Middle East.

From 1990 to 2011, the US and coalition forces expended monumental efforts and monies, but tragically many soldiers lost their lives or were wounded. We must measure the cost against the gain and wonder of the wisdom and value of continuing to allow any terrorist dictator to operate an oil rich country. History has proven the direction of a civilized world is headed for jeopardy with any country's dictator in control.

World Warming

Throughout the history of the world there have been many occasions where warmer and cooler climates prevailed as compared to the present mild temperature clime. Many ice ages have occurred and, according to scientists, the earth is still in the last Ice Age, which started 3 million years ago. The last glaciation peaked about 20,000 years ago with the ice diminishing around 13,500 hundred years ago. Data supports the possibilities of two world glaciations in earth's entire history with methane providing the added atmospheric heating to break the world ice cubes. Humans had nothing to do with the heating or cooling. Many natural complex issues determined the climates.

There exists little doubt humans have added gaseous emissions and co_2 to the earth's atmosphere, whereby US and other countries have instigated a world cleanup program. Although the earth is technically still in the last Ice Age with mild temperatures, little warming has been noted over the last one hundred years. Basically, political leaders want to trade dollars for a temperature drop. From several consensuses of climatologists and scientists, curtailing co_2 emissions on a worldwide basis will have little to no effect on decreasing temperature over a long period. Maximum effort and money expenditures might provide a decrease of .34 of a degree centigrade in a hundred years. Let it be noted that a consensus of anything is not a scientific investigation, however, science does verify the compilation of scientific data. Over many years there has been no real corroborative correlation between co_2 and temperature.

Most informed citizens of the world would agree that better stewardship of the world is required. The methodology is in question. Politically oriented leaders need to understand the environmental problems confronting them. Before mankind, the earth produced 4.65 billion years of varied environment, and during those years, life with propagation began. We do not have a clear picture of prehistoric times, and only the sciences allowed any knowledge of happenings in the broad expanse of time. However, a look at volcano, Toba, which has had super volcanic activity 840,000, 700,000, and 75,000 years ago, presented an environmental problem that changed the destiny of mankind. Toba is still active, but the eruption at 75,000 years ago changed the world's climate and narrowed the world population to less than 10,000 individuals, maybe as few as 2,000. From this Toba apocalypse, mankind bottlenecked, thereby ensuring DNA commonality of 99.9 percent. As a singular volcano, Toba's power and earth changing abilities were and are outside of modern man's measure. Throughout world history there have been thousands of volcanoes, and currently there are 1,500 active ones, not counting those on ocean floors.

Of course, there are many factors controlling climate and warming: ocean currents, tectonic plates, earthquakes, volcanoes, mountain ranges, and natural methane releases are but a few. Political rhetoric for monies to combat global warming seems diminutive against natural forces, which mankind has never controlled and likely never will.

Due to unsubstantiated political rhetoric, a few misnomers, and climate myths, past and present climatological facts are in need of clarification. Presently, there are a few politicians and environmentalists who claim the earth suffers from more heat now than in the last ten centuries. The records prove the idea false. The Medieval Period of one thousand years ago was warmer than today. The area around London could produce fruits no longer possible in today's climate. For the last interglacial period of 100,000 to around 150,000 years ago, a science survey indicated temperatures five degrees centigrade hotter than today. Sea levels were much higher, attributable to less glaciation. At the time, humans were not capable of producing mass volumes of co_2 with flint, a little wood, and cooking fires.

Today, there flourishes an idea of global warming with only a slight increase in the co_2 levels. The idea has no basis. Science has verified that no earth temperature runaway condition has existed, aside from an asteroid striking the earth. Even with co_2 levels in the several thousand parts per million, the earth experienced no runaway warming. Ice cores have given evidence where CO_2 levels have risen after temperature increases, which gives evidence that higher CO_2 levels are not fundamental to the temperature rise. Evaluation of the ice cores have shown cold periods to be drier and windier, as opposed to claimed droughts and floods during global warming. Still, there is a cry that global warming is unprecedented, which also disregards the high temperature rate increase in the pre-depression and depression years. These years were before the populace could produce large volumes of CO_2.

Most scientists believe global warming is from natural causes, rather than human sources. There is no doubt human produced CO_2 is evident but not a prime cause to the warming. The current global warming mania wants trillions of misdirected dollars for CO_2 reduction, and, of course, the US will suffer the greatest expenditure and financial burden. Before the US and other countries formulate plans and acquire funds, scientifically directed panels should debate evidence, histories, and facts. Politicians should be ancillary, not directive, to the panels. Currently, politicians, influenced by consensus opinions, have proposed vast dollars to form a program against global warming, not knowing such monies will attract individuals with little or no interest in a global warming solution.

From humankind's developed rationale, one must wonder why humans are so persistent at destroying forests and vegetation which by photosynthesis yield life-giving oxygen? Better yet, forests use sun, water, and carbon dioxide for food to produce oxygen. Politicians and scientists are disputing the long-term effects of excessive carbon dioxide. A general agreement exists that too much is a detriment to Earth's environment. How much is too much? But we continue depleting forests, our oxygen making capabilities, which depletes carbon dioxide—with an inexplicable rationale. Currently, forests and vegetation are usually replaced by apartment houses and parking lots, allowing more people to produce more carbon dioxide. Science indicates it took eras of time or millions of years to develop forests, which had slow, progressive growth, but humankind's lack of environmental rationale is quickly destroying what took millions of years to create. In the human destruction of Earth's forests, industry and science must form agreements against mindless harvesting of Earth's lifeblood forests.

A simple fact should not be overlooked: neither humans nor money can control the climate or the world. Why is not a more obvious solution proposed? A far less expensive method of controlling CO_2 levels and pollution is to control population growth. But there is a loggerhead among governments, politicians, religious leaders, and various religious beliefs. The US is in an easy example for looking at population growth. In 1950, the US population was around 150 million, but in 2000, the population neared 275 million. Of note, the total world population of around 7 billion required all of human history to produce. The US population could reach 550 million in sixty-five years or less. The numbers cannot be precise, but the US is looking at exponential population growth, which the nation and world cannot tolerate. If faster growing nations are added to this tally, then the problem is obvious and nearly insurmountable. Fighting global warming is not a single solution problem, but scientists and leaders must first identify the contributing factors. Although sticky politically and religiously—the world's growing population is a prime factor.

Since countries insist on individual courses of action for solutions to global warming, international communication and cooperation has been and will be difficult. Countries that give no concern to warming are of the most concern. For long term world survival, international relations and communication must improve, while solutions for human contributions to the warming problem must be meaningful, possible, and reasonable. Survivability is the prime motive, but monies are always of concern.

Long-term survivability forecasts are going to require compromise on all facets of life: government, families, religious beliefs, law, ethics, and corporate operations. Civil chaos is not the goal, but world maintenance will require thinking directed towards cleanliness, rather than satisfying ourselves or organizations we may represent. In any human endeavors, we must realize our combined environmental efforts are pale against natural earth occurrences such as volcanoes, earthquakes, floods, or arriving celestial bodies, as exemplified by the Cretaceous dinosaur extinction 65 million years ago. In the perceived co_2 battle, let not the over enthusiastic environmentalists and uninformed, money hungry politicians forget that plant life uses co_2 as food, while producing the o_2 and the evolved, protective o_3 we humans love so much.

Gun Control—Second Amendment

From the times of the colonies, the Revolutionary War, and finalization of the Constitution, the American nation has been fundamentally based on freedoms, personal freedoms, and the pursuit of capitalistic ventures to advance the nation. The last 250 years have seen the advancement of factors furthering the American way of life: technology, transportation, communication, healthcare, energy, housing, and, at least, semi-guaranteed retirement benefits, to name a few. The human engineered banking and monetary system requires freedoms with trust and a large input of government fiduciary responsibility.

Our forefathers were endowed with greater intelligence and wisdom than some politicians ascribe to them. Some of their acute foresight was undoubtedly accrued from the Revolutionary War and the reasons for fighting it. In our supposedly modern times, America has generated some politicians who have dated and assigned obsolescence to our founders' wisdom. The auspices, technologies, and freedoms of modern times have blinded them to the solid, insightful wisdom inculcated in the Constitution. Unless guarded in their thoughts, politicians can be swept up by emotions of the possibly uninformed masses, modern sentiments, popular fads, and, of course, ideas that keep them in office. Sometimes the issues that maintain the sovereignty of a nation evade politicians' clear thinking and are replaced by popular, emotionally laced concepts, especially those which maintain an office in Washington.

We have forgotten the fundamentals that made this country great and allowed a quick rise to world prominence while other countries with much longer histories remained third world or mediocre in power. What was it that made America great? Initially, we broke away from Great Britain and started with some poorly equipped Pilgrims, who had some basic ideas of freedom and self-governance. The British did not agree with the colonists' self-governing and, of course, the idea was not "politically correct." But, the idea of personal freedoms seemed to prevail in breaking free from the British, and the concept continued making America the greatest nation in the world. Do our current political leaders not realize what made America accelerate beyond other countries? For some, the truth and concept of personal freedoms escapes them. The idea and concept of freedom has not gone away, but for others, emotions and "politically correct thinking" have displaced thinking about national sovereignty. Maybe, the likeness of displacing 100,000 people for one person that is "offended" gives an idea. The fashionable idea of changing governance or redirecting the world for benefit of one or a few socially skewed individuals has got to stop. The thoughts of a sovereign nation must prevail in leaders' minds. Broad, long-term thinking, rather than the current, fashionable, re-electable, short-term thinking, is the necessary cognizance required.

Many modern-day political leaders have overlooked our initial, formative wisdom. The Constitutional Second Amendment is based on personal, law-abiding citizens' freedoms. In the Virginia Ratification on June 5, 1788, Patrick Henry spoke eloquently, "Guard with jealous attention the public liberty. Suspect everyone who approaches that jewel. Unfortunately, nothing will preserve it but downright force. Whenever you give up that freedom, you are inevitably ruined." We should pay heed to this warning, but many socialistic thinking leaders do not, and they are trying to convince others to socialistic ways so they can be sheepishly controlled. To some politicians, a review of our forefathers' quotes on the Second Amendment may be painful and bear some relevant points of truth. In politics, the truth may be painful, hopefully not.

Upon losing personal freedoms, Thomas Jefferson quoted Cesare Beccaria, a criminologist, "The laws that forbid the carrying of arms are laws of such a nature. They disdain only those who are neither inclined nor determined to commit crimes…Such laws make things worse for the assaulted and better for the assailants; they serve rather to encourage than to prevent homicides, for an unarmed man may be attacked with greater confidence than an armed man."

Further, Benjamin Franklin noted, "They that can give up essential liberty to obtain a little temporary safety deserve neither liberty nor safety." There is a great deal of wisdom in these words, and they cry to be applied to some whimsical, emotionally driven political office seekers and current leaders. Instead, we are working under the operatives of "political correctness" and the stagnation of the masses for the perceived offended individual.

George Mason said, "To disarm the people is the most effective way to enslave them." Wisdom overflows from these words. They say a disarmed national citizenry suffers the loss of freedoms and the free pursuit of happiness. Only those who wish to install a dictatorship with a socialistic form of government desire disarmament of its citizenry. Why? A dictator fears an armed citizenry. History has proven the relationship, but the learning curve on the issue is incredibly low.

One more quote from Sam Adams: "The Constitution shall never be construed to prevent the people of the United States who are peaceable citizens from keeping their own arms." Sam Adams and many others had wisdom well beyond their years in time. Yet, we have politicians who would dismantle the Constitution and the Second Amendment. The reasons and answers are contained in the concept of control, and the obvious fact that a dictatorship cannot rule over armed and free people.

Why is there so much rhetoric on gun control in recent years? Part of the answer lies in the number of people who want control of the citizenry and the government. Obviously control is a direct infringement on personal freedoms. The government agencies that have come into existence in the last twenty-five years smacks of a quest for more government control of all life aspects. As noted by wise leaders of the past, less federal government is exceedingly better than more government. In short, the US government is too large and has too much control.

Free citizens of a sovereign country with a sound, workable Constitution such as America's, need to guard against those who would trash the Constitution to install a more socialistic and dictatorial form of government. Through the coolness and misdirection of oration that leans heavily on "give me" and "I deserve" programs, socialistic programs arise that tend toward governmentally controlled citizens and gun confiscated existences with loss of all capitalistic ideals and personal freedoms. Citizens should be aware of prophets offering something for nothing. The thought begs for increased education. Never has there been something for nothing. The trap—there is more to gun control than gun control, and the far-reaching consequences are adversarial to personal freedoms.

How Large—the Government?

At some point, one must notice the number and size of federal agencies. Older data shows a federal government employees headcount at over 2.7 million. The tally using federal, state, and local employees under a full or part-time program is over 22 million. That means the ratio of government employees to the total population is high, expensive, and unacceptable. It is a given the country needs employees, but the questions are how many employees and how many government agencies are functionally necessary? The government has no definitive answer on the agency count, but an estimate is between 700 and 800 agencies. Is such a large number of agencies necessary? One must use a large imagination to quantify the monies to operate the agencies, let alone the salaries and retirement benefits of employees. The government is the major contributor to the growing national debt. Such squandered waste is scandalous and indicative of fiscal irresponsibility.

What is the need for Homeland Security to have twenty agencies and over sixty different offices? From our colonial and post-Revolutionary War days, we have advanced a great deal in inefficiency and major expenditures in a government for the supposed benefit of the people. The taxpayer must support fifteen different departments under the Executive Branch, and a labyrinth of bureaucratic tiers, which by the sheer number of offices can't foster communication or efficiency.

Under the United States Department of Education, there are approximately forty offices, and while the US has the best colleges and universities in the world, the country suffers from a national decline in secondary education. Decline has been prevalent for some years and evidenced by a lowering of ACT and college entrance exam scores. A decline in number of high school students bound for colleges or universities is noticeable in many states. For some high school graduates, remedial courses on the basics are necessary.

An assumption that secondary students are less capable or intelligent now than in previous decades does not seem prudent. However, an idea that they are receiving less than an adequate secondary education might be warranted. There may not be a definitive or singular answer, but some problems may be in the concepts of what is "socially acceptable" or "politically correct." Promoting students to remain in the social or peer group, while they are deficient in the core curriculum of their grade is detracting from college longevity and adding to admission problems. Population growth and overcrowded classrooms are contributing factors to the workloads of teachers. From the early days of secondary education, a high school graduate was educationally equipped to deal with most life and employment requirements. Currently, less than a full scale of high school graduates shows capabilities of suitably educated citizens. Without a doubt, there are intelligent, educated secondary students, but, sadly, less that in the decades of the past.

Every department of the federal government cannot be detailed in description, but under the Department of the Treasury, the Internal Revenue Service is worthy of comment. Some notice must be directed toward the 100,000 employees that require a yearly budget of at least $11.2 billion. One question must be asked: If most US taxpayers are honest and will pay their share of the national income, why do we need the heavily funded and overpopulated IRS? It seems a fair question. From a layman's point of view, we do not need the IRS—except for special tax provisions. The numbers of provisions and tax codes are nearly without number, and the federal government has allowed this proliferation. Exemplary within the vast amount of codes or loopholes is the emergence of noncitizens or undocumented workers who generate large, unwarranted refunds through false claims of non-present children. Such claims siphon off large amounts of money, which the deficit can ill afford.

As in the precedence covering the last 200 years, Congress keeps a diminutive eye on the military, especially after a war or military action such as World War I, World War II, Korea, or Vietnam. Monies not spent on the military can be redirected to favorite spots of prominent Congress leaders. This tactic works well for reelection campaigns but jeopardizes the country if any part of the defense organization needs to function immediately with efficiency and positive results. Almost continuously, any branch of the armed forces needs a financial boost. History teaches that emergencies requiring military forces happen and then military might be generated to deal with the crisis, which is inevitably late. With the post-crisis generation scenario, monies can be redirected, but in the interim generation, the nation is defensively weak and vulnerable.

When looking at the size of the federal government, the debt structure must be examined as well. The US has powerful statesmen, representatives, and senators who are not concerned enough about the national deficit. Fiscal irresponsibility with a direct deficit pushing 20 trillion dollars indicate, at the least, a laissez-faire attitude with the debt crisis. A statesman might ask, "Do we have a debt crisis"? Therein lies the problem. For our future and survivability, a great country cannot tolerate the discredited and shameful posture of a runaway, uncontrolled debt structure. Looking at the future with a tremendous debt, forecasts some undesirable possibilities for the nation. An inability or unwillingness to meaningfully address the debt invites socialism and a nondemocratic form of government. There are stronger words for the possibilities of future rule, but the nation and leaders must do more than add rhetorical financial rules and rationalization against the monstrous debt problem. Unless directly addressed, the debt problem will not go away. So far we have only danced around the financial problem solutions, and some leaders give the appearance of ignoring the problem might make it go away. Maybe, addressing a nasty national problem is not fashionable for well ensconced government leaders, when the answers are going to be politically and socially painful.

The US is at or near where the Romans were in terms of government size and the amount of unsolved problems, especially a future financial conundrum. The US is at a crossroads; which road will we take? The hard road leads to capitalistic freedom and sovereignty, while the easy road leads to socialism with the losses of identity and freedom. The losses result from debt default and hopeful forgiveness, the keys to a mundane existence and third world, nonsovereign, stature.

DOD Budget

Great countries or empires that come into existence and subsequently disappear harbor many reasons for failure: immorality, leader aggrandizement, cronyism within leaders, or leaders bent on acquiring wealth rather than establishing worthy leadership roles. Statesmen's leadership roles must place their country first. Countries that did not establish and enforce definite moral leadership rules, such as the Roman Empire, were doomed within a few hundred years. Eventually, Rome could not defend itself due to moral issues and the inability to find a protectorate army or navy, which provided greatness in its formative years. America has been great for 250 years for reasons of a free people willing to maintain freedom with strong defense forces. In those 250 years, America has become more united and has made more sacrifices to freedom fighting forces than those threatening its freedom.

Over the years, the defense budget has grown steadily except for the heavy increases in war and conflict. The wars in Iraq and Afghanistan were financial peaks, and by the middle of 2011, the US spent $1.3 trillion. The arguments for the worth of those sacrifices, military personnel wounded or killed, and monies spent are without end; the US maintained the budgetary finances as necessary as in other wars and conflicts. In a review of history, the one fact that has kept many enemies from marauding our borders and shores has been the overpowering might of our military forces. These forces are comprised of specialty organizations of the Army, Air Force, Navy, and Coast Guard—with some overlapping duties. From an outside view of any of these service branches, an enemy of freedom and capitalism has a formidable task of forcibly conquering the US.

From within, however, we must be cautious of misdirecting priorities. In 2014 there were plans to reduce the defense budget by billions of dollars. Civilian defense leaders usually start the easy processes of limiting healthcare benefits, pay raises, housing allowances, and commissary benefits for the military. In some areas, lower ranking personnel qualify for welfare, food stamps, or both. Such a situation should not exist in the military. The easy thinking toward expenditure cuts targets a reduction in force, RIF. At certain times, the RIF has eliminated the most highly experienced officers and enlisted personnel. Many of the RIF personnel were the higher paid cadre. As they left their services, they took the experience and the expertise with them, which the services badly needed. From outside influences and expenditure pressures, defense leaders sometimes equate efficiency with dollars saved from reduced benefits and personnel. With the reductions went capabilities as well. In all matters of reductions, the Joint Chiefs should be consulted—and heeded.

In times of peace and prosperity, congressman and senators observe monies going to the defense budget with the desire of diverting funds to special interest projects. Congress controls the purse strings, but the Secretary of Defense would more aptly control diversion of funds. The US has made diversionary mistakes before, usually after a war or conflict, and as the next defense crisis arises, the military is behind in the required effort. Politicians, appointees, and leaders newly elected to offices must guard against the ever popular military budgetary monies diversionary tactic for close to home projects.

As with other government departments and agencies, the DOD has added many agencies to those of earlier years, and the matters of capability and efficiency are questionable. With more agencies come more people and with the greater personnel count, more leadership arrives. The questionable growth under the auspices of "need" requires a bigger budget for what is deemed efficient. The US has three branches of service and a Coast Guard, but under the DOD there are more than seventy agencies or entities. These kinds of numbers are getting inordinate and must be questioned. The leadership salaries for these entities question reasonableness.

The DOD has over 800,000 civilian employees with another three quarters of a million contract employees. The number of nearly 1.6 million civilian employees is greater than active duty personnel. No matter what the rhetoric is from, politicians, leaders, or secretaries, the civilian employee count clearly shows an out of control condition. Civilians don't fight the wars, soldiers do. We have a problem of recognizing priorities at the higher levels of leadership. As Secretary of Defense, Leon Panetta, said before Congress, "Frankly, I don't think you should de-trigger sequester on the backs of our civilian workforce." In a further statement, Panetta testified, "I mean, I realize that savings could be achieved there, but the civilian workforce does perform a very important role for us in terms of support."

Continued DOD operations with misunderstood priorities will place the DOD at a tremendous operational and financial disadvantage. The functionality of defense will be severely impaired, regardless of the monies directed to it. As leadership cuts the numbers of active-duty soldiers, the Pentagon civilian workforce has grown by 13 percent. Reasons for the civilian workforce increase from the DOD have not been straightforward. Yet, according to the Congressional Budget Office, CBO, the civilian pay will be two thirds of the DOD operations and maintenance spending for the time period 2013-2021. Since the DOD does not have a database for civilian contractors, work and costs are difficult to control. In 2010, the Government Accountability Office could not produce an audit opinion of the US government. The principal obstacle was "serious financial management problems at the DOD that made its financial statements un-auditable." In 2011, seven DOD entities received unqualified audit opinions. Is anybody watching the store?

As with many complex problems, there is no single solution, but a realignment of priorities is a standout for a beginning point. We must do a better job of caring for existing military personnel. They must have sufficient pay to live in our society, health benefits, commissary, housing, education, exchange, and other necessities must be maintained or increased. However, the benefits must not be counted as pay. For longevity in the forces, family services are necessary. For stability and predictability, there must be a curtailment of the oscillating service member numbers, requiring DOD to reach an operational number for military personnel and maintain it for each branch. The President, Congress, the Joint Chiefs, and DOD should sanctify the number. Conflicts can increase the number, but the services should not go below their operational number, as congress and or senators find reasons to divert monies.

A few politicians have requested new jet fighters for the Navy, which is a positive move forward. Naval histories of the US and that of Great Britain, for example, have given the directing country commanding advantages and positions throughout the world. Since World War I, carriers have given the Navy the precedence for victory, bringing the modern aircraft carrier to a power with which warring entities must contend. The fighter aircraft aboard the carriers must be equal or, preferably, superior to the task, populate the carrier with more than adequate numbers, and be able to survive the predicted target hostility. A multibillion dollar jet fighter is not needed to eradicate a few guerrillas hiding behind a sand dune. For the Navy, carrier fighters are extremely important, and the priority of the carriers cannot be shunned. Naval admirals are seeking a sustainable fleet of more than ten super carriers. If budgets hold, two more are under construction, and one more is in the planning stages. However, fulfillment for the Navy requires a total ship count of over 300.

For the DOD budget, quickly acquiescing to the term "obsolete" can put undue stress on appropriations and manufacturing. Obsolescence of equipment may originate from a government contract manufacturer who no longer can easily or profitably acquire materials or make the parts. Contractors and the DOD certainly want aircraft capable of task multiplicity, provide a lengthy service life, and be amenable to modifications or future upgrades. In the past, however, DOD has acquired aircraft with a supposed multiplicity of capabilities, and with this plethora of capabilities, the aircraft were, at best, mediocre in task accomplishments. The assumed savings and manufacturing money and time for the extra capabilities imposed overruns in both money and time. From a pilot point of view, an aircraft well suited to fewer tasks is better than mediocrity in many tasks. The defense dollar goes farther with superior accomplishments. The obvious coordination between DOD and manufacturing is critical and mandatory.

The savings of DOD money for replacement of the B-52 bomber has been a sterling example. The early models came into being in 1952 and became operational in 1955. The bomber has been in existence for over sixty years, and in the cadre of aircraft, models A through H, various records were established. The longevity of the B-52 has been its modern design, resilience, and ruggedness. Over its service life, the variety of models have accommodated modifications for better capabilities or longevity. All models were capable of high subsonic speeds with operating ranges beyond other bombers. The refueling capability was, indeed, a requirement, but for many assignments, the B-52 could reach its target unrefueled. On long-range assignments, subsequent modern bombers could not reach the operational area unrefueled. If the tanker failed, so did the new, expensive bomber. Meanwhile, the B-52 flew as required. With care, the remaining B-52Hs can do another twenty-five years.

In 2002, DOD was considering a re-engine program for the B-52H. Had the program been sanctified and accomplished within a reasonable time, the money saved could have been in the billions. But, indecision entered the picture in the early 2000's budgeting and pricing. For an airplane bound to operations until 2040, it makes financial and operational sense to re-engine the B-52H, as its current engines begin to show signs of added maintenance and inefficiencies. Since the initial re-engine consideration, many years have passed, but the B-52 continues to be a viable, reliable weapon system. The re-engine program still looks good. There are many variables in cost and expected long-term savings. The front runner advantages would be increased range on an already range devouring aircraft, greater loiter time, less maintenance, and less operational fuel costs. Reliability would be enhanced on aircraft that is one of the Air Force's most stalwart aircraft. The B-52 has set records in range and endurance; however, a further range or endurance increase would eliminate refueling tanker expenses on some missions. Tanker fuel and maintenance expenses were not calculated on B-52 re-engine costs since they are quite variable, but if a tanker is required, the costs are large. Old numbers show a price of four billion dollars for fleet re-engining, and the old data yields a savings of 4 million dollars per bomber. Still, in dollars saved, the pro-factors are great, and the tanker factor is not considered.

From the initial design, the A-10 Thunderbolt II or Warthog has performed with excellence. But from DOD and higher, the thoughts of expenses and obsolescence are being directed toward the aircraft. Had the A-10 been advanced to earlier production, it would have been ideal in Vietnam. The Warthog was fast enough for expediency, could decelerate for target precision, loiter on target for effectiveness, carry adequate munitions, endure ground fire punishment, or utilize its 30mm cannon to dissuade just about any enemy aggression. Field repairs have been standard operational maintenance procedures, increasing utility and reliability. The A-10 is super redundant in many functions, including engines, hydraulics, and flight controls. For such capability the DOD paid an initial price just short of 19 million dollars per aircraft in 1972.

The aircraft the A-10 was replacing, directly or indirectly, was the venerable, time proven A-1 Skyraider of World War II, Korea, and Vietnam.

Secondarily, the A-10 replaced the Vought A-7D Corsair II. The A-7 had a most admirable war record in Vietnam with its bombing and gun accuracy unsurpassed. The A-10 was designed and manufactured for close air support, and although its predecessors were excellent, the Warthog is unsurpassed in its close air support role, especially for its price.

While the A-10 was introduced in the 1972, its first combat role was in the Gulf War in 1991 and subsequent Middle East conflicts. As expected, the Hog turned in outstanding performances. The DOD and the USAF were getting the biggest bang for the buck. In the Gulf War, the Warthog flew over 8,000 sorties with a launch capable rate of nearly 96 percent. It is hard to ask more of an aircraft which is subject to battle damage.

Although Congress and DOD have raised expenditure questions on A-10 refurbishment, Boeing was awarded a contract of over 200 million dollars for redesigned and more efficient wings. Such a rejuvenation would extend the life of the A-10 to around 2035. Additionally, the A-10 has been amenable to navigation and targeting upgrades since inception. Conservation of monies while increasing capabilities on existing aircraft should not be confused with settling for mediocrity. For the skirmishes and conflicts in the current world situation, an upgraded version of the Warthog presents itself as an ideal aircraft. Of course, replacement aircraft are coming, but billions of dollars to cover what millions can do is not cost effective in the interim.

National Debt

There are financial gurus in the US who proclaim the US debt to tax revenue is acceptable for the time being. With an understanding that modern times add toleration to indebtedness for the furtherance of economic growth, a nation functioning with an ever increasing debt has little tolerance for emergencies. Soon the US direct national debt will reach 20 trillion dollars, while the tax revenue is well over $3.5 trillion. The future holds a guarantee each of these numbers will increase. Generally, the US populace is not terribly concerned about the debt because the US Congress is falsely not worried. The nation must be concerned about the cost of debt, which, currently is nearly 242 billion dollars and increasing. Our saving grace is that US GDP, gross domestic product, is well over 18 trillion dollars. Maintaining this number or increasing it, is a good idea. While the total US debt is nearly 65 trillion dollars, US unfunded liabilities are nearing 102 trillion dollars. Of course, these numbers are increasing.

Some financial experts feel it is fine to live with debt in order to heighten the economy in the wake of projected debt increases over 150 percent of GDP. In short, trashing America's future for a current increase in economy is not prudent. To maintain the present debt to GDP ratio, thinking appears confined to cutting spending or raising taxes. From a citizenry point of view, raising taxes has a limited horizon. However, cutting federal expenditures has many possibilities with areas which the public knows little. For a comparison on financial health, no corporation or private enterprise could survive with the debt structure such as America's, and the country can legally print more money—inflation. A review of history reveals no country has spent its way out of debt, which some politicians think is a financial plan.

While it is necessary to point out that presidents, senators, and congressman should be thinking in terms of national greatness and sovereignty—thinking in terms of pet projects, constituent desires, geographically limited projects, or party needs have driven federal expenditures to unimagined heights. At least financially, leaders need to forecast and think on a national scale. Why is the nation working at nearly a 750 billion dollar trade deficit, which is increasing? Are we a little soft on foreign trade policies? In this case and to strengthen the nation internally, thought must be turned to external trades. Improving the foreign trade deficit requires concentration on both exports and imports, and to gain profitability, strong personal negotiations are necessities. Here are excellent opportunities for the Executive Branch and Congress to back the Secretary of State in trade negotiations. Congress must consider why so many corporations place manufacturing and labor in foreign countries. Would a more conducive tax structure help? It might help the economy with more workers at home paying taxes.

America became a manufacturing giant in the late 1800's and continued through the 1900's. Freedom, capitalism, and energy were the driving factors. Complexities and special needs in our society have obscured—for some—America's reason for greatness. To drive the engines of manufacturing and successful enterprises, America utilized the available energy—oil, coal, and natural gas. Primary to manufacturing was coal and the exploration of it. Somewhere in today's "socially correct" society with technologies, leaders have overlooked the driving power to greatness—energy. Correctness, special needs, and environmentalists have acquired a disproportionate voice in asset and energy control. The people who wallow in technological advances failed to realize they are utilizing powers they want to eliminate or diminish. Cronyism, rhetoric, and innuendo obscure the need for new pipelines of oil and gas critical for national growth and economy. A growing economy will combat the national debt. Civilizations cannot exist without regulations set by politicians, but neither regulations nor politicians can power the nation—maybe, by hot air.

Under the Department of Energy, or DOE, the Federal Energy Regulatory Commission, or FERC, the US has approved a few energy corporations to export liquid natural gas. Since the process is complex, safety, regulations, and environmental concerns are tantamount. The US has been burning natural gas since oil exploration began, and there remains vast quantities of natural gas. For this reason and with reasonable export fees paid by the importing nation, continuous deductions from the national debt could be made. Anything the FERC could do to facilitate licensing more exporting companies could alleviate some debt.

Politicians and environmentalists want to put the kibosh on substantial uses of coal, even with coal having been a major player in industrialization in the early 1900's. After Great Britain had its Industrial Revolution, America followed suit and had its Industrial Revolution to become the world's leading industrial nation. By the late 1800's, America's mining industry possessed three quarters of a million miners, whereby the early 1900's produced more than half a billion tons of coal. The nation's early industries were driven by the plethora of fossil fuels—coal and oil. From the late 1800's to the early 1900's, use of coal was rampant. America was built and connected by the ensuing railroads furnishing materials, labor, communication, and people to populate the West. The railroads used coal as did the shipping industry for international commerce. In these times, neither America nor other industrialized nations were concerned about the environmental consequences of coal dust and smoke. As the nation was driving to profits through cheaply fueled mechanization, industries were minimally regulated by the government, causing harm to the environment and population. From those days, technology and skills friendly toward the environment have progressed and promise to improve.

In the US, coal still furnishes about 50 percent of the fuel to produce electricity. Even environmentalists like the convenience of electricity, but an unknown time stands between coal pollutants now and what would be ideal—for sulfur dioxide, nitrogen oxide, and carbon dioxide removal from coal burning. Environmentally acceptable exhaust gases from coal burning is in the future and is determined by the speed of technology. The DOE has held carrots before industries using coal, and considering recent technological advances, a genuine coal burning and cleaning process should be close at hand. Incidentally, the US landed a spacecraft and men on the moon with far less technological advances than currently exist. With a US patented coal cleaning process, the country could export coal and the process to third world countries with fees for each, and thereby drive down the national debt.

Technology is advancing on oil extraction from shale, which is plentiful in the US. The extraction of oil from shale could assist in driving down the national debt. With vast quantities of shale available, oil companies are working on better extraction technologies. Presently extraction is possible, but the amount of energy required is high. A technological success will help the national debt. By sheer indigenous quantities, iron is already on the export list, but, perhaps the US is not receiving satisfactory payment. An indicator of lagging profitability is the US shipping industry, which exports iron for the mining industry.

The US has exported technology for years as verified by the tremendous profitability of the Middle East oil export industry, especially Saudi Arabia. In retrospect, the US should have gained a greater degree in profits for expertise provided and in override agreements for oil production—forever. Without a doubt, the US and other countries should work diligently on decreasing pollutants, especially carbon dioxide. For those developing the technology, the cost must be passed along. By natural endowment, the US has many resources to export. Unfortunately, many of them pollute, and to export fairly, the US must find means or ways to decrease or eliminate byproduct pollutants. Retrospect shows the US has the potential capabilities, but financial and environmental initiatives need to be applied. For US national debt assistance, nearly pollutant free resource exports could help.

US Oil Reserves

For the last twenty-five years or longer, many sensationalists and environmentalists have touted an end to world oil reserves in the near future. In the same time, technology has advanced oil recovery procedures and given hope to oil production from shale, with the largest supply being in the US. Major oil companies are researching shale oil production, and to date oil production from shale is possible but not financially feasible—yet. The correct methodology is being pursued, but with the scientific efforts of major oil companies aligned, humans will prevail. The estimate of shale oil production is difficult and at some variance, but the states of Colorado, Wyoming, Utah, and Montana offer shale oil production hope. Including the Colorado Green River formation, the total reserves could reach over 3 trillion barrels of oil with a recovery number of 1.5 trillion barrels of oil. That recovery number is greater than reserves of Saudi Arabia and many other oil producing countries combined. With such a vast energy reserve, the US and the world could have a new lease on a bright human existence.

Oil companies are on the verge of a major oil and energy breakthrough, but solar and wind power industries are receiving nearly all the news for their modest energy contributions. Solar and wind energy efforts are noteworthy, since mankind must continuously seek energy alternatives, but they might not exist if politicians had not provided generous subsidies. An optimistic approach to oil exploration combined with better recovery techniques and the tremendous shale oil potential could give the US and world several hundred years of fossil energy. Regardless, alternative energy research must be continuous.

Environmentalists are not in a minority nor alone in the fight for a cleaner, garbage free world. Currently, nearly everyone, corporations—including oil companies—want to keep the world free of pollutants. By nature and life, species do pollute, but rational, long living species will want to keep their home, the world, as clean as possible. Unfortunately, some humans become emotional and overzealous for a goal and want to reverse engineer some life enhancing, life-giving ideas or products. Interestingly, some overzealous environmentalists strike or protest against the products or procedures they enjoy and from which they reap benefits—unknowingly. Oil companies are firmly entrenched on environmental cleanliness or face heavy fines, and fines are not their cleanup motive. They want the environment clean too. But in many civilized advancements, electricity, steam machines, railroads, ships, aircraft, and interstate highways, the environment was dirtied in the development and advancement stages. The process is part of civilization and ongoing life. Of course, leaving a mess is not excusable or friendly to civilization or the environment.

Let us not forget the existing oil reserves, not shale, the US and world have been tapping for 150 years. There remains oil in the proven locations, but the easy production has been taken. Yet with advanced technologies, maybe 60 % of the remaining oil could be obtained. Overzealous use of some recovery techniques, acid fracking for example, has caused concern for freshwater aquifers, but other techniques such as co_2, water, steam injection, or other recovery methods are much closer to being environmentally friendly.

Alaska has proven itself lucrative in oil production, although the winters and lack of northern road systems make drilling operations difficult. Since 1977, Prudhoe Bay has supplied the US with millions of barrels of oil; still, there remains much of northern Alaska open for oil exploration. Why haven't we explored these regions, except for one test well—with secret results? ANWAR, Arctic National Wildlife Animal Reserve, might supply oil for several years, but Congress has put the area off limits. It is a primitive animal refuge and provides a rather large area for wild animals to live. All environmentalists want to preserve the world in a pristine state, and those of us with any appreciation for the world can understand world preservation. However, as people and countries of the world, we must coexist to survive. To be sure, oil exploration has harmed the world and residents of the concerned areas, but in recent years with renewed efforts toward regulation compliance, top flight oil companies have and will make efforts with expenditures for not harming the world. A simple observation is—for now, or future times, the world must continue and so must life—for all species. Problems of the here and now are essentially infinitesimal compared to energy and civilization of the future. It would seem energy and civilization are directly proportional. Less energy means less prosperity, productivity, and fewer peaceful civilizations. History might indicate that the major belligerents of World War II, Germany and Japan, were fighting for oil and energy. Have we forgotten the lesson? Do we currently have better energy sources than fossil fuels?

A good percentage of individuals pledging their allegiance to the environmentalists' cause may be misinformed about the operations and workings of society. The underlining motive should be the ascension of world societies to a cleaner, more livable world. The attainment of a pristine world is probably not possible. From the cadre of well-intentioned people joining the searching ranks of environmental nirvana, emotion could supply a disproportionate share of the motive, not reason. Environmentalists are certainly not the enemies of the world, but their joint actions could be misdirected and provide a hindrance to civilized progress. The federal and state governments have more than sufficient regulations and laws to protect the environment. In fact, the laws may be too stringent. The problem then, is not insufficient thinking along protection of the environment but the lack of enforcement applied to industries that have a potential for messing on the world. Environmental folks must realize that the great civilized country of the US, for example, rests on and was derived from the bastions of industries like mining, shipping, refining, transportation, manufacturing, grocers, agriculture, and many more. In the process of becoming the greatest country on earth with free people, some environmental problems arose, and now there are alternatives and laws for protection. Some politicians have aligned themselves with a specific environmental cause, which is fine, but politicians should function better in the government arena.

From growth, progress, and advancement of civilization, clearly, the US and the world needs to expand oil exploration. Humanity must admit to a lack of motivation for energy, and the fossil fuel of oil remains the supplemental medicine until humans become more innovative. A concern for the environment is necessary, but we must use caution not to become counterproductive. As energy advances, in any form, the world will advance accordingly.

WORLD GOVERNMENT

Throughout the history of civilizations, there was talk and concern about a government of the world. Civilization is young in the world and those governing it are inexperienced. Wars and conflicts from the ancient past to those of the present lay heavy evidence that humanity is not ready to proceed under world government. Intercountry conflicts and revolutions, world wars over control, supremacy, energy, and wealth are strong indicators that humans have not arrived at a point in civilized thinking for world rule. Many countries, including the US, have politicians who believe a unitary government is the answer for mankind. Such a government is not yet the answer—maybe, in many centuries when wisdom is plentiful.

Why is it that the world contains many countries, languages, customs, and unlimited forms of government? Simply, people want autonomy in their lifestyles with a country reflecting their societal wishes. The country borders are there for a reason, and from time to time countries fight over them. Sometimes the battles are over territory, but societal ideals, rights, and freedoms are motives of conflict.

The last 5,000 years contain many rulers, kings, and emperors who had quests for control over people and as much territory as possible. Ancient rulers did not know how much territory existed, or they would have tried to rule it. While technology has advanced to assist modern leaders, the world still has leaders whose thoughts differ little from that of greedy emperors many thousands of years ago. Their rule over the proletariat would still be archaic. For now, leaders in a world government environment would be tempted by excess power and opportunities for greed. The factor of greed may arise from the disproportionate relationship of few leaders compared to world rule. Presently, leaders or statesmen do not appear to possess the wisdom or knowledge for such of an expansive rule.

For world government, the coexistence of many religions will present problems demanding vast amounts of leadership time and concern. Here, the wisdom may lie in the knowledge that there is no solution. While the mainstream of religions is peaceful enough, there are subsets of religions that are quite eager to proliferate war for control. To a degree, these radical subsets are territorial, but modern amenities, such as air transportation, allow the radicalism to spread. A world government, rather than a strong local government, may not be able to directly deal with religious conflicts or wars.

Using the US as an example, communication in the same language, occasionally, has resulted in – no communication. By extending communication problems of one language to a world rule with hundreds of languages would tend to foster problems. The UN has done well dealing with lack of communication, but that is a full-time job ensuring the true meaning of communication—difficulties and problems abound.

Living under a world government does not automatically mean life could proceed without a military force. But whose or what military force would be dominant? A Unitarian force might be tried, but it would lack the forcefulness and precision of a "homeland" directed force. The inherent problems are communication and standardization; these lead to eventual ineffectiveness, which would indicate a failing world government. Should a unified military force be generated, as the world contains an excess of "qualified" admirals and generals? Each has been inculcated to their own country's systems and cultures. They are products of their societal rules, customs, laws, and standards, which can make it difficult to psychologically attain the demeanor for correctly commanding a world military force. From this aspect, world government would have a higher probability for failure.

For world government, prudence should be exercised for countries' technological compatibilities. As can be imagined, some countries are well advanced on technologies and have many intra-system controls by computer. Transportation control, ground and air, might be greatly assisted by integrated computers, while satellites could control advanced agrarian applications. First rate hospitals have advanced computer systems for research, analysis, and patient maintenance. However, third world countries possess little or none of the computer or technological advantages, or if the country does have technologies, it is reserved and restricted to the ruling class. In comparing and integrating technologies, impossible is not an operative word, but advancing third world countries to a good technological footing would present a monumental and costly project. For some years a world government may not have the manpower or financial abilities to advance backward countries.

A successful world government must address, solve, and operate a workable financial system, which, at the minimum, would include a universal denomination with banking and taxing systems. In the world there are many denominations of currency, and whatever the number, financial gurus are going to spend time on equivalency. Under world government, countries must be on a par for exchanges, trades, imports, exports, investments, banking, and taxation. For such actions to occur, a major meeting of the minds that operates central banking must be successful and, to date, such successes have been rare or nonexistent. The purposes of joint meetings vary, but in recent history, World War I, World War II, Korea, Vietnam, and the Middle East exemplify failed meetings of some highly qualified leaders.

Technically advanced countries should have few problems establishing inter-banking systems, and, in fact, many international banking systems have been operational for years. Third world countries or those with limited or no fiduciary systems will have problems joining a modern, international banking system. Credibility will be the operative word for backward countries. As can be seen, fiscal minds will need to educate the third world countries on banking, credibility, transfers, and exchanges. The task will be monumental.

On a world financial basis, a basic question must be asked: Will some form of capitalism still exist and will investing still be a possible financial endeavor? Initially, if investment opportunities are lost, then capitalism and civilized advancement will grind to a halt. If world government and rule calls for a cessation of advancement in the financial world and provides no reason for personal incentives—then the system will stagnate and degenerate to a socialistic world. The world already has this type of non-contributing society that provides the sinkhole from which an advanced, capitalistic society must overcome.

Addressing taxation under a world government will provide problems, requiring the best financial minds. The plethora of US tax codes would be of little use, and they exist for special tax circumstances, which, for the most part, should be eliminated. Establishing an equal footing for presently existing countries for taxation under world government will be a formidable task. World government leaders will be tempted to bring third world countries more in line with advanced countries for trade and financial exchanges. To do this some unequal tax advantages are likely to occur. If advantages exist, then the world government will have retrograded to socialism, allowing loses of all perceived advantages of world government. After analysis, a flat tax will probably be the tax calculator—long overdue.

Only a few hundred years ago, the Pilgrims departed England to seek personal, political, and religious freedoms, which were being usurped in Britain. These people were willing to sacrifice everything to venture to a strange, unknown world to exercise their freedoms. Presently, it seems incredulous that leaders have emerged who believe world government is humanity's answer. Under world government, the freedoms the Pilgrims so bravely and ardently pursued are very likely to erode when coupled to world rule. While the world has not gotten larger, the world population has expanded tremendously. World rule will most likely reverse the freedom and individuality process. Again, rulers and countries need more thought and useful incentives for curtailing endless population growth, rather than a world rule.

SOCIALISM

Every so often there are calls in the US for a more socialistic form of government and way of life. More equality is touted for the middle class and poor. Wealth, possessions, or both should be redistributed among the less fortunate by a disproportionate tax on the rich or those who excel to prosperity. Disregarding inflation, there are calls for the minimum wage to be raised. Accordingly, the price of goods and services will rise to a level to nullify the minimum wage increase. If socialistic aspirations reach fulfillment, the state will administer all medical issues of the people, resulting in the usual, historically mediocre care. The rich or industrious will, by means of taxes, carry the load of a free university education for all students. So far, in a successful, capitalistic world there are no free rides. In the pure socialistic system, manufacturing, sales, and wages become uniform whereby equality and uniformity are stressed, maintained, and are goals of the government.

Late instigators of socialism were Marx, Lenin, and Stalin. Properties were held in common and prices of commodities were in proportion to production labor plus a profit to the state for replacement of commodities. Wages were for subsistence. Socialism is not a new concept; it has been around for thousands of years. Why? It plays on each generation's gullibility with the concept of something for nothing and security within the state. What the system belies are two classes of people: the elite or those in control, and the proletariat or the uninformed, ignorant workers. Many countries and systems have adhered to socialism for a while, but the system breaks down as eventually there is no money, materials, or substance from which to draw.

History has shown the elite or upper-class in numerous countries have touted and wanted a socialistic form of government. The forms of socialism are without number, but all tend to favor a working class with subsistence wages distributed evenly to the proletariat, working members, of the society. Ownership of properties is not condoned. From the disciplines of finance and science, nothing is free, although the uninformed in societies, including the US, believe that governments can offer something for nothing. For the socialistic government to function, the working class must be returned to a monetary value less than the equivalent value of their labor. The difference or profit must be used to support the government and the ruling or elite class. The incentives for the working class is a status quo life and government security. Medical care is offered and supported by the government, partially supported by a high tax levy. For the elite, there are early, generous pension plans of which the proletariat are blissfully ignorant.

While socialism has had many names and nomenclatures, the well-traveled road of history is consistent in portraying socialism as an elite controlled government to which the laboring masses pay homage. High taxes are a step in the route to socialism, and inevitably the taxes are paid by the working class. Yet, the socialistic governments can show no correlation between high taxes and a decrease in poverty. The idea of increasing the taxes on the rich for a rise in the level of living for the poor or poverty class doesn't work. A higher tax for the rich can work in reverse. Greed of the elite in the working government will foster the idea: if a little tax rate is good, a higher rate will be better for all concerned, including the rich. A breaking point will be realized when the rich cease investments that produce money. They will be unable to return money to industry and profit making endeavors. At this point, everyone suffers, including the government. A closed system with zero incentives for the masses, who are encumbered with high taxes so the ruling class can operate the government, is doomed to failure. Open and free markets have success and failure, but the free market is enabled for a recovery. The socialistic system has little means of recovery; taxes for the impoverished working-class must continue—failure.

Pundits for the success of present day socialism give examples of Scandinavian countries. These countries tout the auspices of socialism by having high taxes and welfare systems to carry individuals through life. A closer look at the five Scandinavian countries reveals a heavy dependency on capitalist markets.

While liberals and socialists point to Norway as an exemplary socialistic state, Norway is very successful because of its capitalistic interests in oil production and is a large exporter of oil for Western Europe. Scandinavian countries have a tendency toward high taxes and cost of living expenses, while the US has the most disposable income compared to Scandinavia. As demonstrated, high taxes in a socialistic system does little to help the lower or working-class, while economic prosperity catalyzed by a capitalistic system will help the lower classes. Because of small and cohesive nature of Scandinavian populations and their capitalistic enterprises, these countries are prosperous but not because they favor of socialism and high taxes.

For their premier countries of socialism, a measure of satisfaction of the citizens might be indicative of socialistic return to its society and proponents, but aesthetic factors are difficult to measure. If a comparison of Scandinavian suicide rates is measured with other countries, including the US, American rates are found to be much lower. Such a comparison is not conclusive for the success or failure of socialism, but in a true socialistic state, the inability for enterprising individuals to rise above the masses by recognition of profits through entrepreneurship becomes frustrating. Advancements through individual efforts are nonexistent, and any such technological or humanitarian advances are more likely to evolve from capitalist origins. True socialist and communist states in their developmental stages needed suppression from capitalist news telling of technological or humanitarian progress, because of the demoralizing effect on the working, subjugated proletariat. Existing as a citizen in the proletariat is normal and a way of life. In socialism and communism, a rule of equality is broken—for the controlling government, ruling class, and elite to continue in existence, the working class must labor at non-equitable rates and render taxes. Proletariat incentives are not given, and the laborers must maintain a status quo.

A kind of predatory socialism works in modern times. Some politicians advocate socialism offering to tax the wealthy for redistribution of free perks to the middle class and the poor. A precursor to this flawed theory is that it overlooks how the wealthy acquired wealth. Of course, capitalism was the means to finance private industry, which produced goods, services, and re-investable profits, leading to progress. The current in vogue socialism wants to pirate the wealth of the upper class rich for redistribution, disregarding the capitalism initially creating the wealth and how this idea will vanquish the money sources. This type of socialistic thinking is good only for the here and now and those in the uninformed "give me" class.

When socialism is scrutinized, it has built-in obsolescence as indicated by its historical failure in countries touting it. Excess monies are absorbed by the elite while little is a returned for manufacturing. Value of goods and services is attached to labor involved. Since no capitalistic exchange exists, replenishment of materials is difficult or nonexistent, and a continuum of operations may cease.

For less-developed countries, dictators may embrace the idea of socialism, since it offers subsistence for the masses and exists in the here and now. Little thought is required, and ingenuity or entrepreneurship are not involved. Outstanding technological or cultural advances are not likely to occur, nor are continuation monies. With the ease of socialistic instigation comes dictatorial control, obedience, and conformity of the masses, since the state must reinforce socialistic and obedient behaviors.

Within socialism, individuals are kept from entrepreneurship and enthusiasm by the meager state subsistence and the demand for obedience. From socialized societies there comes few inventions or advancements for civilization. In fact, many socialistic states have received aid from capitalistic countries—mostly the US. Some monitors of foreign aid contend that free money promotes socialism, and a government that manages the aid money is practicing socialism. While many US citizens believe the percentage of the yearly national foreign aid budget is quite high, the actual percentage is around one percent, yielding 35 billion dollars of foreign aid to over 140 countries. Without a doubt, some of the monies go to socialistic countries. In Africa, humanitarian needs take precedence over political guidance causing the continent to receive a major share of the total aid.

In South America and Bolivia, US foreign aid money established some advantages of proto-democracy. Bolivians and socialist president Evo Morales enjoyed a few of the advantages of democracy. However, the citizens took advantage of the democratic ways and were demanding more from the basically socialistic government headed by Morales. Moving toward democracy was not to happen. President Morales, basically a dictator, certainly was not going to be removed from power; so, an alternative was a rewrite of the Bolivian Constitution whereby Morales accumulated more power. For example, he had an unlimited term in office and the judiciary branch was formed to his favor.

Under these circumstances, his socialistic ideals could prevail, and he could not be deposed. Has the US given Bolivia foreign aid monies? Yes, in 2013 the US gave Bolivia nearly 700 million dollars. Similar actions have happened in other countries, but any political failure was due to internal strife or corruption, not US intervention.

Unless word from the outside world is completely blocked, socialistic or communistic citizens generally seek personal freedoms that perpetuate a capitalistic society. For some years, foreign aid monies from the US have directly or inadvertently helped socialized citizens seek freedoms not allowed within their government. As can be reasoned, the US cannot indefinitely fund the aid to countries where civil freedoms do not exist. At some point, the oppressed citizens will need to initiate movements for personal rights. The US cannot save or police the world.

A lengthy history of socialism has produced failing systems. Communism, a socialist system backed by a strong military, has failed. Russia is a mere shadow of its former self. The world has recorded massive failures of the socialist or collectivist systems, which are pocked with starvation and poverty in the working classes. Why is this? Humans are more intelligent than the socialist system allows. Many citizens, not all, want to rise above the masses or excel in some manner. Eventually, enterprising citizens will crawl out of their government controlled confines and begin realizing personal aspirations are only limited by their imaginations, if living in a free, capitalist society. Countries espousing such freedoms are great. Failing socialist societies indicate a serious deficiency of humanitarian learning and understanding.

Fresh Water

For thousands of years humans have been wasting precious water. Cultures and growing populations were the driving forces to increase agricultural production and avoid hunger problems. Certainly, in earlier times agrarians must have clung to the idea of an infinite supply of fresh water; the importance to the farmer was maximizing crop production. Any method for getting water to the field was satisfactory. Efficiency or conservation of water was not honored, while food production was rewarded in profits and by the recipient society. Parts of the Middle East give testimony of freshwater overuse, particularly, Iran and the United Arab Emirates.

As the world population increases, the demand for food crops will grow in a commensurate manner. But the supply of fresh water for irrigation and civilization needs, such as sewage disposal, is quickly diminishing. There are no more freshwater aquifers from which to draw and irresponsibly waste water. To solve the world shortage of freshwater, humans must first recognize the water shortage problem, and, to date, the required admission of a water shortage is not apparent. Therein, the problem festers, and as world societies function, problem-solving does not occur until a crisis is at hand. The US is no exception to the delayed behavior.

As humans continue with the wasteful use of water, the worldwide situation will inevitably produce political strife and conflicts. To avoid adversarial situations, humans must communicate and cooperate to form joint-effort based solutions. Humans, by wars and conflicts, have proven an innate inability to cooperate, forecasting the delay in addressing water conservation. But time is critical for the conservation problem, and the scarcity of freshwater must be brought to the forefront. Most of us know the world is 71 percent covered by water, and about 97 percent of that is saltwater. Already a saltwater to freshwater conversion has been attempted, but the expense and energy required are far from practical. Somewhere between two and three percent of the water remaining is fresh, but much of this water is in ice caps. Approximately one percent of the world's fresh water is available for human use. We do not have as much consumable water as once believed. So what is civilization doing with its freshwater? Usage is tied to irrigation, municipalities, households, and industrial uses. Of these uses, crop irrigation is by far the heaviest consumer. Much of the runoff water from these necessary uses is, unfortunately, nonconsumptive, contaminated water. In the inefficiency of the high percentage run off, there may be a conservation solution.

In world areas where the water shortage is desperate, a serious consideration of recycling sewage waste water is in contention, or the procedure has been accomplished. Contaminated drinking water is responsible for killing several million people a year. Cholera, parasites, and dysentery are but a few of the maladies caused by contaminated water.

Industry probably falls into one of two categories of water conservation. Either the industry expends funds and efforts on water conservation, while being responsible to its community, or the industry cares little about conservation and water cleanliness to maximize profits. Industry has the means and technology for conservation and cleaning of water, but from CEOs to stockholders, profitability is a major factor. This is an area where industry leaders and stockholders must realize the battle for freshwater is here and now. The crisis is at hand, and the alligators are in the water. One can't drink profitability.

Household water waste is quite evident. Over-irrigation of lawns and gardens, long showers, running sink faucets for no particular reason, washing machines with small loads, car washing, or playing in sprinklers are but a few household water wastes. A conscious effort to conserve would help. Without doubt, food crop irrigation must be accomplished to feed the ever-growing world population. Agrarian science must introduce efficiency to irrigation such that more crops are watered with conservation in mind. Farmers must irrigate more crops with less water, which means there should be little to no wasted runoff water. Spraying water gets the job done, but encourages water loss by evaporation. Running water in ditches or canals also encourages evaporation, or any open means to transport water is higher on evaporation. Investigative hydrology is not proposed, but runoff water presents sure signs of wasted water.

For irrigation, drinking quality water should never be used. Freshwater from sources of rivers, lakes, streams, recycled, or shallow aquifers should be used first. Deep, fossil water aquifers should be used as a last resort and certainly not to combat a drought. There is finality in this maneuver. For preservation of humanity, deep aquifers should be reserved for emergency drinking only. But with drought, farmers have quickly tapped the precious fossil aquifers, which rarely replenish themselves as they were sourced by ancient glaciers or other prehistoric sources. By depleting the deep aquifers, we are sealing our fate for short term gratification and profits.

Possibly, crop selection should be considered for decreased water requirements versus crop maturity and earlier consumption dates. With monitors, data, and satellites now available, farmers are closer to crop and water efficiency than previous years. From efficient, low evaporation sprinklers to water injection, farmers can avoid water runoff and adapt a crop to the field and water potential. In areas forecast for low water availability, rice, a high water demand crop, would be at a disadvantage. But to utilize the field and increase water efficiency, corn, southern peas, sweet potatoes, watermelon, beets, and other less water demanding crops can be grown, and with no runoff, efficiency further increases. For example, California grows rice; yet, the state has been desperate for water conservation for years. Money and profits are factors, but they will not be if freshwater runs out.

The most obvious, yet undesirable, solution to water conservation is worldwide cooperation on population control or growth. The most dominant parameter within which we must live is the closed water system of the earth. In other words, the water volume of the earth has been constant for several million years, and outside of a catastrophe, there should be no change. By 2050, the world population is forecast to be 9.5 billion, possibly higher by independent studies. World population by 2100 is estimated to be 11.2 billion, about 4.0 billion more than present. If estimates are correct, world hunger may also be on the increase. Civilized societies are continually adding buildings and parking lots where crops could be produced. With the added 4.0 billion people, waste water and garbage increases in a logarithmic manner, and more importantly, consumption of freshwater increases tremendously—and from where?

As country populations continue growing, especially in Africa, and droughts materialize, farmers are forced to draw freshwater from the deep, fossil aquifers. In the future, how will humanity solve the impending water shortage crisis? Are we to run the train until there are no tracks? We must solve the problem. As in many complex problems, the answers are complex as well. Technology can provide much assistance with conservation and can help clean some water, but consumption and waste by ever growing societies are serious problems seeking timely solutions. In time, our consumption rate will exceed both rate and quantity available. Are we aiding our own droughts, famines, diseases, pestilence, and possible extinction?

To face the consumption part of the problem, civilizations will need to face two variable components and one constant in the problem. The variables are governments and religions, while the constant is the earth's supply of freshwater. Governments must agree to try for solutions. More difficult, religions must cooperate in some manner. There is no longer a need for large families in the world. In the distant past for agrarian or feudal war reasons, large families were necessary for survival. Modern times have reversed the large family penchant, due to mechanization and technological reasons. Steadily, various religions have touted large families to increase members and spread the advantages of the specific religion. If such philosophies continue increasing the world population, a crisis in water consumption is inevitable.

If religions believing in large families do not understand or cooperate, the dreaded loggerhead of state and religion may occur, and certainly no one seeks such a situation. A prediction of outcome is difficult, maybe impossible, but a large degree of survivability is at stake for everyone. Continuous religious and state communication, understanding, and cooperation is essential. Given enough time, the conservation/consumption problem will demand action toward a solution. If otherwise, our future may be nothing short of chaotic.

Learning from History

Has the world or the US learned from history? From societies and the humanitarian progress of nations, Homo sapiens possess little learning from history's teachings. The wars countries have initiated serve as examples. Participation as a warrior in the cause of a country is a noble thing. But wars for the personal causes of leaders or politicians are counter to rational thinking. Leaders, presidents, admirals, and generals must lead, guide, or dictate from reliable intelligence agencies' operations. Never should emotion have a part in the decisions of war making. Unfortunately, many past battles have been fought on the emotions of leaders. Germany's motives for Nazism, fascism, and the desire for world empire came from egotistical emotion and the need for war making materials.

Did the world and the US have warning of Hitler's rise to power and his motives for world empire? From the early 1930's to 1940, his ambitions were known by Great Britain and the US. The world and the Allies were comfortable in their complacency, especially around September 1, 1939 when Hitler attacked Poland with 1.5 million troops and the Luftwaffe. Simultaneously, German U-boats attacked Polish naval forces. Hitler argued to the world that such actions were defensive. In the months before Germany's attack, Winston Churchill had vehemently warned Britain of Hitler's untrustworthy, aggressive nature. Hitler was threatening Europe with fallacious oration and armament. The British Parliament was contemptuous, showing complete disregard of Churchill's warning. Before the attack on Poland, Neville Chamberlain had recommended the Parliament go on recess from August 4, 1939 to October 3, 1939. With Europe in a turmoil—caused by Hitler—wisdom did not prevail. The learning from a lesson was available, but Britain, and the world chose to ignore it, except for Churchill. In a surprise and confounded effort, Britain declared war on Germany—again—on September 3, 1939. The lesson of forgetting the treachery of Germany added to complacency was devastating; yet, the Fascists were arming well beyond limits in the Treaty of Versailles. For some reason, no one noticed Germany had exceeded the World War I treaty on arms limitations. The warning was evident, but Britain and the Allies were enthralled with complacency.

His brusque, factual approach made the truth hurt, since the British still held Churchill in disdain, but during the late 30's, Churchill was espousing his acid learning from bitter history. On August 4, 1939, Churchill warned the US about Hitler's conquest of Austria and Czechoslovakia, while Mussolini worked his acts of freedom. History adamantly records that Germany has started two world wars and bears watching today. However, pointing out the instigations of two world wars and keeping a watchful eye on Germany is not considered "politically correct" nowadays.

It is not that Germany and Japan did not provide plenty of prewar evidence; these fascist countries were quite generous with Empire warnings—even vociferous. Since the closing of World War I with Germany under the Treaty of Versailles, the Allies assumed a placated nature, a concern for internal affairs, and desires for individual comfort. Prewar concerns and initiatives to prevent a war required national concerns followed by efforts. The lack of action on the parts of the US and Britain probably comes from—it can't happen again—attitudes. Depending upon whose measure of time since World War I, there were only twenty-three years until the US entry in World War II. The span of time was certainly enough for comfort and complacency to enter mainstream thought. The semi-surprise of the Japanese attack on Pearl Harbor provided the US and Allies the fighting will to attack Germany and Japan.

Prior to Britain's entry to World War II on September 3, 1939, three days after Germany's attack on Poland, the Prime Minister, Neville Chamberlain, had tried appeasement with Hitler. Chamberlain gained no concessions for Great Britain and was deceived by Hitler's attitude on coexistence within Europe or the world. Unknown to Chamberlain, Hitler saw the world, or part of it, under the rule of the Third Reich. Hitler was an excellent orator and actor, duping Chamberlain by his eloquence of coexistence and nonaggression. The realization of what Hitler said and his intentions were entirely different concepts which Chamberlain did not fathom. As the Prime Minister, Chamberlain reported to Parliament his impressions—not as aggressive as perceived—of Hitler. Unfortunately, Chamberlain's impression of Hitler was grossly misleading. Hitler was willing and capable of killing millions of people, if they were not Aryan and he needed their territory. Apparently, Parliament was deceived as well, forming status quo modes of operations—until September 1, 1939 and the German takeover of Poland. The lack of learning and undetectable deceit nearly put the kibosh on Britain and the Allies.

After Hitler's attack on Poland in 1939, the US abstained from entering the war even when several friendly European countries were in the balance of being conquered by crawling Nazi fascism. US leadership words of wisdom were those of maintaining protection in isolationism. The world is large but small in politics and threats of war. In 1939, war in Europe had begun, and the US maintained friendly, cooperative relations with European countries. Worse yet, the US saw beginnings of U-boat interference to international shipping along its east coast waterways—torpedoed ships. Undoubtedly, US leadership understood the inevitability of war and hoped for its confinement to Europe, but in the interim, leaders tried for life as usual—until Pearl Harbor. The status quo words of isolation and derived comfort were suddenly obsolete. The US and leaders were not learning because the country was in a dangerous state of denial.

If countries and civilizations are not learning from a myriad of past mistakes, where do major universities and colleges stand? A generalization cannot be made, but a historical event at the Oxford Union in 1933 prompted a consensus student statement: "That this house refuses in any circumstances to fight for King and country." This squalid, shameless certification was at a ratio of nearly 2 to 5. According to Churchill, the shameful message was sent by the young at Oxford University—Britain's most famous University. Although Britain tried to downplay the Oxford student consensus, Churchill stated, "It is a very disquieting and disgusting symptom." For a sovereign, mighty nation, such a pacifist message gave a doubtful and shameful view of Britain's future leaders. Churchill was so frustrated that he issued, "Mankind is unteachable. They refused to learn. As a result, history keeps repeating itself in endless catastrophes."

Of course, the question arises as to how the Oxford students came to the pacifist, defenseless attitude. Churchill aimed the arrows of cause directly at Oxford professors. For the time, Churchill accused educators of wanting to teach, but they also needed to learn in order to pass on the understanding implied in wisdom. World War I had been the largest manmade disaster in the world, and educators had not learned from it.

In the 1930's and prior to World War II, higher education in the US was extensive with colleges and universities, but they were more concerned with industry, production of corporate captains, and a more educated workforce. In the early 30's—from a US point of view—Hitler was a dictator with which Europe could deal. As 1939 approached, Hitler was becoming a big wolf and was ready to gobble up some territory. Still, the US view of Hitler's fascism was at length, "Europe could handle the problematic war." This was a prominent point of view—while US isolationism lasted. As Hitler began pressing the world for a fascist empire, the US population and college age people continued their attitude of noninvolvement. Although World War I was an ideal example for learning, the US was having none of it.

A big leap in thought: had Britain and the US learned from World War I and understood the treachery of Hitler before September 1, 1939, could World War II have been prevented? The same question applies to Japan. As factors for inaction, 1939 was only twenty-one years from the close of World War I, and the belief that another world war was brewing—especially instigated by recently conquered Poland—seemed imprudent for isolationists. From a European observation, Germany was easily exceeding the limits of the Versailles Treaty, but any alarm went unheeded due to a disregarding of historical learning lessons.

Moving up a few years whereby World War II can be viewed in retrospect, we can note it was the largest trashing mankind has given the world. From World War I to World War II, what has civilization learned to avoid World War III? Not much. Ask the nearest twelfth grader or college student about the wars or reasons for them. Let us grant that professors have gained some wisdom on the world battles. There is no guarantee.

More than a third of a century ago the world and the US began having serious, enduring problems with Iran. 1979 saw Shah Reza Pahlavi forced out of office by Ayatollah Ruhollah Khomeini, the Iranian religious leader. Khomeini asserted Shah Pahlavi had too many western associations, which were direct confrontations to Khomeini's Shiite founding and devout studies of the Muslim Quran. Thereafter, under the guidance of Ayatollah Khomeini, Iran became the number one safe haven and proponent of terrorists. Although Khomeini did not appear to engineer the student takeover of the US Embassy hostages, he seized the opportunity to demonstrate Iranian hatred for the US. For the US lack of learning, sixty-five embassy hostages paid the price of 444 days in Iranian incarceration. The US military rescue debacle further punished the US and was exemplary to the lack of learning and leadership necessitated by past historical treacheries.

As history moves forward a few years, Iran must be bribed or paid not to pursue nuclear weapon technology—excellent statesmanship. With little doubt, Iran wants to be recognized as a sovereign country. Some years of research have proven 3.5 percent enriched uranium was feasible for Iran, and the plutonium program was advancing as well. Iran had already enriched uranium to 20 percent and was looking for further increases. 90 percent enrichment is the doorway to nuclear weapons, but the US, UN, and other nations don't want an Iranian 90 percent uranium enrichment program. With underground and hidden facilities, super enrichment might have already been possible. The US and UN do not know. Basically, the consensus was not to trust Iran; therefore, the sanction on trade and oil export was put in place. Iran had earned the no trust factor.

Now, with inspections by the UN nuclear watchdog, sanctions are to be lifted, and some of Iran's frozen assets are to be returned in a pro-rated fashion. As understood, the monies are Iranian assets, not those of the US or UN. The sanctions have impoverished Iran, and, maybe, a return to commerce and profitability will convince Iranian leadership that Middle East coexistence without nuclear weapons is possible. However, such talk is merely ideology to adjacent countries such as Saudi Arabia, Israel, United Arab Emirates, Oman, Bahrain, Qatar, and Iraq. The repatriation of monies may be 150 billion dollars. It is not clear how much money has been frozen, but the assets will not be returned at one time, hopefully.

The past and the lowest order of espionage should have revealed Iranian untrustworthiness and unwillingness to live under basic UN treaty rules. Simply, Iran is not dependable for forming international relations. Since Ayatollah Khomeini's days of rule, Iran has become more divergent from the norm in world societies and is now home for Islamic radicals. Syria operates as a satellite state to Iran and also rates very low in trustworthiness. Maybe, sanctions should not be lifted, but if they are, a very slow, non-trusting, heavily inspected UN guided program should be in place. If areas are off-limits, then sanctions should remain.

FOOD WASTE

In the face of gross waste of water and food, one must wonder the fate and destiny of an ever increasing world population? Humans have an innate propensity to procrastinate problem-solving to the point of working in a crisis. To humanities' credit and intelligence, we have learned how to produce sufficient food with technology and mechanization, but we have 800 million people starving in the world while there are nearly 3 trillion pounds of food that never get consumed. What a conundrum. Worse, as world civilizations move toward the mid-twenty-first century and beyond, food waste will magnify unless serious conservation measures are taken.

Food waste would feed twice the 800 million hungry if wasted food could be efficiently redirected, packaged, refrigerated, and transported to populations in need. But without reason, these and other food conservation actions are not performed to a significant degree. Why are these needed actions not occurring? The world wastes about 30 percent of food produced, as fruits and vegetables have the highest potential for becoming refuse. Developed countries consumers have become spoiled. For produce markup, grocers have taken the tact of dressing up produce as if it were to be showcased. Food that is good but does not have the classical look or has a small blemish is not displayed but instead discarded. The food is still fresh, nourishing, and completely edible but not beautiful. Beautiful fruits and veggies fetch a better price for the grocer and cover the price of the discards.

Developing countries waste produce by means of inadequate refrigeration, improper storage, and transportation of food from grower to grocer. However, customer appeal is still important. Although some loss of produce may occur at the grocer, underdeveloped country customers may be more interested in quantity per price than beautiful produce. Bruised fruits and vegetables lose importance versus edibility—a bad trait.

Wasted food is an environmental loss. In the US 30 to 35 million tons are thrown away each year, while one in six Americans are hungry. But unconsumed food is not the only loss. Losses of well over 150 billion dollars are suffered in the food industry. Markup prices can cover grocer losses, but producers are fighting the profit war. Considering worldwide declines of water, fuel, seeds, pesticides, and other growing requirements, waste is inevitable. Water, of course, is a major, valuable resource to lose—enough to fill a major river.

Upscale restaurants go for eye appeal since high paying customers demand the best, but in the appeal there are many discards for color, shape, misplaced tips or ends, and temperature blemishes. Prices must cover the discards. In the US, schools waste considerably more than one third of the food served. Mostly, the waste emanates from set portions, which are not entirely consumed. Maybe, a better way is to allow students to select portion sizes they want. If tables of food were set and all sanitary conditions applied, unused food could either be stored or given to charitable institutions, increasing food usage.

At some point in the food business, producers need some leeway in dealing with grocers on deliveries deemed substandard. In fact, early communication would be ideal for anticipated produce deficiencies. Acceptance of a lower contract price might work. Of course, spoiled food is off the table, but blemishes, off-color, misshaped, or different ends does not detract from nourishment. Grocers that specialize in downgraded produce are ideal for conservation, while the food has the identical food value as the aesthetically appealing produce. Such food could go to charities or institutions. Possibly, producers could sell downgraded produce to institutions or canning companies for large productions of soup.

As a byproduct of the world's wasted food, over a third of the greenhouse gas is produced. The food wasted is not necessarily spoiled. It may not have met the customer specifications, the quantity of food was more than required, the shape or color may have been slightly different than normal, the end of the selling day was near, or food to the garbage heap was the convenient disposal method because proper storage and refrigeration were not at hand. In many restaurants, second day food may not be allowed, especially by upscale hotels, eateries, clubs, caterers, or exclusive diners. These businesses would be embarrassed and receive a downgraded reputation if their estimated food quantity was low when preparing for high profile luncheon or dinner occasions. It is deemed far better for business to throw excess food away, rather than explain the shortage to high paying customers who expect excess and opulence. They are paying for it. Unfortunately, they are paying for more than waste in the discarded food scenario.

By 2050 or 2100, what will be the food-waste-hunger situation? Without some basic changes in the producer-grocer-consumer food sequence, the world and the US will have some major hunger problems. 2050 will produce a world population of 9.0 or 9.5 billion people, which will require an increase of 70 percent, probably more, in food production. The freshwater withdrawal for the agriculture to support the increasing growth is phenomenal, as well as the supporting assets of land, labor, seeds, etc. If present population growth rates continue, will the year 2100 have the necessary supply of freshwater?

How is the world to solve the impending food problem? Fresh water conservation and a cessation of food waste should begin at all levels: producers, transport, grocers, restaurants, institutions, and the home. Consumers, businesses, and individuals need to lower specifications for appeal of produce while becoming more accurate on estimates of requirements. Even spoiled food can be used for other than consumption. Smaller portions of food served at large institutions could save waste. Very soon, civilizations need to realize the importance of land for agriculture. For high profits, we use acres of land for parking lots adjacent to businesses or apartments. Of course, the apartments are for the ever increasing population, which is becoming a problem. We need to watch and prepare for our future destination—hunger, poverty, or survival.

Common Core Education

For many years, the US secondary educational system has experienced a continual decline in the academic abilities of the average high school graduate. This fact has been verified by comparisons of college-level entrance exams of US students to foreign students. Remediation for college aspiring high school graduates is considerably higher for the US students, with language and mathematics being the areas of weakness.

The weaknesses can be independently verified, but how is it that the US secondary school system has consistently produced less rigorously prepared academic graduates in the last several years? As in many complex issues, there is not one all-encompassing answer. In the last twenty-five years or so, a few social rules have crept into our behavior and customs. One rule of social behavior is that of being and acting "politically correct", and from that behavior, we have derived what is "socially acceptable." There are many facets to these behaviors, but we have evolved efforts now, to not offend secondary education students. Some factors of the so-called "repressive" behaviors are: to be in second or third place, to lose a basketball game by any number of points, to receive a lesser academic score or grade, or, simply, not be the winner. In some secondary school systems, leaders and teachers have founded a "do not offend" approach to teaching. In essence, no one fails due to lack of academic rigor.

Highly educated people, although not necessarily educators, produced the Common Core Standards through a quorum of twenty-four people. There was a well-intentioned effort in the Common Core Standards. Essentially, common core focuses on mathematics and the English language. Of course, the common core curriculum must be developed by the state, district, and leaders to reach the Common Core Standards. In the minds of those designing the Common Core Stock Standards, CCSS, the old educational system was failing, regardless of any socially acceptable attitudes applied. Was the old system really failing? Or, were the standards a little high for leaving no child behind?

When applying new parameters and rules to any system, difficulties inevitably follow. But in the application of common core over a system which has been tested and in place for 100 years, there were objections from students, teachers, educators, and parents. A perceived symptom of the common core system was that of dumbing down, as reported by parents. Methodologies for working math problems are far more important than deriving the correct answer. Under the old system, producing the correct answer was indicative of a correct methodology. In attempts to assist their children, parents were occasionally at losses for methodologies or reasons to problem solutions. Of course, the parents were products of the previous older system, but when describing, delineating, or graphically depicting the math problems, there is some visualization of the problem, eliciting a better understanding. Although memorization of the times tables is necessary, memorization of abstract formulas can be disregarded.

Common core has brought some ancillary, needless, freedom-depriving baggage. With the government subsidies, the taxpayer is responsible for affording dossiers on students and teachers, a freedom-depriving maneuver. The source of the idea exists but is unknown. With the government subsidies, dossiers on students became necessary. Progress or lack of it must be documented. The dossiers are not private but open to federal inspection. Here, the common core curriculum is not causing the rift of privacy, but the regulations attached to the government stipends do present an invasion of privacy.

In framing the Constitution, the federal government was purposely not charged with mandatory administration of education, since the forefathers saw wisdom in the states providing the educational function. The states are best suited for exercising these rights of freedom in education. For some years, the states and the federal government have watched and suffered from a decline in college or university entrance exam scores by high school graduates. Even students with high grade point averages were scoring lower and being assigned to remedial classes in math or English to gain full university privileges.

The plunging academic scores were certainly noticed by the government education department and individuals, whether in the education location or not. For the sake of the nation and America's future, something must be done to increase academic rigor. An enigma ensued: should the older methods of education continue with enhancement or should a newer method of education be tried, which emphasizes conceptualization rather than rote memorization of formulas and arithmetic methodology for problem solving?

From the government, states, and many concerned individuals, the decaying levels of academic achievement for secondary education students were an endangerment for the nation and our future. The declining academic achievement has been a factor, and a solution is needed. In finding a solution of presenting standards at the grade levels, the National Governors Association, NGA, sponsored an initiative and formed a group of well-educated and educationally inclined individuals to develop mathematics and literary standards. The developed initiative was to: "provide a consistent, clear understanding of what students are expected to learn, so teachers and parents know what they need to do to help them." To abide by this statement, problems easily arise on knowing what to do for helping students. The standards developed are to be robust, collegiately acceptable, and meet career or work standards.

People in the NGA and the Counsel of Chief State School Officers, CCSSO, have labored extensively over the Common Core Standards, but implementation and methodology of meeting the standards at each grade level are at the discretion of the states. There are many arguments for and against common core standards. The standards are not field-tested, and core standards have not been in use long enough to give valid results. A couple of states have reported only modest increases in test results, and equalization of state academic achievement has not materialized. While thirty-three states have accepted Common Core Standards, educators and parents complain they do not want the federal government interfering with the states' jobs of secondary education. A further grievance for the states was the tremendous cost of implementation, and if the state accepted federal monies, there were federal criteria with which the state had to comply. Here, there is a problem of the federal government dictating to the states.

Are there advantages to the Common Core Standards? The nation and states recognize that academic standards have declined, and some positive, corrective action must be implemented. Educational analysts have indicated the

Common Core Standards are superior to the majority of states in mathematics and English. Easily, standards can be written to a higher level, but achieving them with a good methodology is a different matter. How did the states evolve to such lower standards of academic rigor? A few individuals in industries see a worthwhile effort in raising academic standards, especially industries involved in technology. And from a university's viewpoint, higher academic achievements and acceptable entrance exam scores make orientation and beginning class assignments more meaningful.

What happened to the old methods of teaching? At one time some years ago, there existed academic rigor. The methods of teaching in the early and mid-1900's provided captains of industry, manufacturing, victories in world wars, orbiting satellites, advances in medicine, and missions to the moon. In those years, rigor in the sciences and literary abilities was more than adequate. From the 1980's or thereabouts, the US began the "politically correct" influence with its detrimental side effect of "socially acceptable", which became embedded in the secondary school system. For reasons of social acceptance, students began seeing a trend of promotions with their peer group, regardless of academic standing—no social traumas allowed. To help students remain with their social group at promotion time, academic standards in mathematics and literacy were accordingly lowered—maybe, to include the class dummy. Athletic competitions and other competitive events began exhibiting "there are no losers" syndromes. Existing in the real world workforce, corporations, and true academia of universities or colleges is a true competitive, capitalistic nature and attitude. If secondary students are receiving outstanding marks and yet are well below college entrance minimums or a failing student is socially promoted, any system in secondary education is going to fail its objective— college or workforce qualified graduates. In previous years the older education worked if students were instilled and rewarded for work, study ethics, responsibility, loyalty, dependability, honesty, and respect for the teachers. In the past few years our more socially correct society has lost many of these values. Reinstating these values and ethics will take some time, as worthwhile efforts take processing time.

The Common Core Standards are having problems as well, and nowhere in the standards are emphases placed on the values of ethics that should be instilled in the successful student. Common Core Standards have considerable thought and work in them, but for success, students must possess and utilize the values mentioned. The nation is spending billions of dollars on the Common Core Standards, which may be successful, but in years well before Common Core Standards, teaching methods worked when coupled to the necessary student values. Once again, we must question our methods to a successful national destiny.

Celestial Havoc

Among astronomers, geologists, physicists, paleontologists, and other scientists, there is little disagreement that the earth has a significant history of asteroid or comet impacts. From astronomers and others, the question for the future is when, not if, another significant impact will occur. Small meteors destroy themselves in the earth's atmosphere daily, but every few hundred, thousand, or million years, asteroids of significance strike the earth. Because of Earth's atmosphere, weather, erosion, and earlier tectonic drifts, many meteor craters have eroded and been lost. However, of geologic recency is Meteor Crater or Barringer Crater, which occurred approximately 50,000 years ago in present Arizona. Worthy of mention is the 4,000 feet wide, nearly 600 feet deep crater that was caused by a 160 feet meteor colliding with the earth in the Pleistocene epoch. In earth time, 50,000 years ago was recent. In the more distant past, but still fairly recent in earth's 4.65-billion-year history, an asteroid collided with the earth about 66 million years ago in what is now Chicxulub, Mexico. Scientists refer to the extraordinary happening as the Cretaceous-Tertiary boundary or the K-T. Others may call the collision The Cretaceous-Paleogene boundary or K-Pg. The asteroid was approximately six miles in diameter and caused a crater of over 100 miles in diameter. Although celestial happenings are not frequent, they have the potential of disaster or life extinction, which happen to a large degree at the K-T boundary. The dinosaurs were wiped out and almost all large sea creatures were eradicated. To human advantage, the dinosaurs are gone, but similar celestial dangers still exist—rarely but surely.

Scientific advances in the last 200 years have given humans, to a small degree, insight on the histories of the earth, our galaxy, and the universe. Our knowledge is feeble and may be likened to studying a grain of sand on the beach through a microscope and guessing at the world structure, but our technology is growing rapidly as is our galactic perspective. We have reason to study earth's earlier histories. From earlier times, we have learned of growth, disaster, and the realization that some events can reoccur. As scientists have learned, some colossal celestial events of the past have brought about extinction of many life forms. A worthy note is: 95 percent of earth's life forms that ever lived on earth are extinct. It would be beneficial for humans to pay some heed and attention to methods for avoidance of celestial catastrophes or extinction events. In the last fifty years, the US and other countries have shown increased interest in learning of past and possible celestial disasters.

While celestial disasters for earth are few compared to earth's 4.65 billion-year history, such catastrophic events have occurred and will again—hopefully, in the distant future. For a little prehistory review, within the Paleozoic Era, 541-252 million years ago, the end of the Ordovician Period, 485-443 million years ago, saw the demise of greater than half of marine invertebrates. Again, the Devonian Period, 419-359 million years ago saw a great end period extinction. The causes of these extinctions are not known, but celestial causes cannot be ruled out or were partial causes of the extinctions.

The extinctions continued with the end of the Permian Period, 299-252 million years ago. With the end of the Permian Period, the Triassic Period, 252-208 million years ago, began, but the greatest dying in earth's history was the Permian-Triassic boundary. In this transition 95 percent of all life on earth died. Since scientists have had the technology, equipment, and broadened earthly education, the P-T extinction was discovered, which initiated the questions of why and how. The P-T extinction was centered at 252 million years ago, but we need more information for world preservation. The likelihood of such an event occurring again is close to zero, but the chance is not zero.

Some scientists, although there exists no complete agreement, subscribe to a probability of sequential catastrophes. Singular or multiple asteroids or bolide hits may have initiated a chain of extinction events, possibly the Siberian Traps, which covered nearly 800,000 square miles of earth with lava. If lava flowed into sea beds, methane hydrate deposits could have released vast amounts of methane. The huge amounts of methane would wreak havoc with all life forms to include extinction. 252 million years is sufficient to hide most evidence of an asteroid collision, especially in ocean beds that subducted their beds every 200 million years or so. Possibly, from a large asteroid, an antipode (opposite on the earth) event occurred, causing additional volcanoes to contribute to the Siberian Trap volcanoes. The P-T extinction is the only known earth catastrophe to destroy all insects, and insects at the end of the Permian were considerably larger than now. After the P-T extinction, some scientists believe the earth needed 30 million years to fully recover flora, fauna, fish, and land animals. Although the Permian-Triassic extinction was, as best we know, the worst extinction the world has known, the Cretaceous- Paleogene, K-Pg, or older term Cretaceous-Tertiary, K-T, was more recent at 66 million years ago.

Modern times and technology utilized by scientists have revealed consistent celestial dangers. While the dangers have not been frequent, the dangers are real, earth changing, and extinction capable. The recent past has seen the US take some countermeasures for NEO, near earth objects, or celestial bodies on a collision orbit, but, presently, we have no spacecraft ready to launch any kind of intercept or counter mission. Our bureaucracy and permission seekers could not launch a counter mission in less than four years. Budgetary concerns have been an independent to countering asteroid intercepting missions, but when the mission is assigned a world or humankind saving event, financial concerns seem rather insignificant. For a closer concern, 99942 Apophis has a very low chance of hitting a return keyhole in 2029 for the 2036 return and remote collision possibility. The biggest celestial problem for earth survival is ignorance of unknown celestial bodies, asteroids or comets. To counter a large asteroid problem, nuclear devices were initially proposed, and soon there was a cry and complaint of nuclear weapons in space. Upon deeper reflection, the nuclear weapons are intended for the preservation of humans and danger of extinction, and, therefore, any weapon of significant destructive power would be satisfactory. By not using nuclear weapons in space to preserve the human race, who are we trying to save—turtles?

Near Earth Asteroids, NEA, are the most prevalent by the watch of NASA, Minor Planet Center, LINEAR, B612 Foundation, Sentinel, Spaceguard Survey, and other agencies. NEA may number between 10,000 and 20,000. Undoubtedly, many more exist. Asteroids and comets of size greater than 140 meters are of concern, but asteroids larger than one kilometer must be found and tracked.

Devices and methods for destroying, orbit altering, or so modifying the NEOs existence such that they are not a threat to earth are limited only by imagination. However, practicality and capability are factors. For destruction or orbital alteration, the nature of the NEO must be known. A nuclear device against the conglomerate of space rubble would probably be a waste of the weapon and delivery system, but such a weapon against a large, solid mass might prove effective. Having enough time for orbit changing by reflected sunlight, lasers, gravitational attraction, or changes by rocket engines could be temporarily beneficial, but in the long term, buying time for a decade or century does not solve the NEO orbit problem. An orbit change only solves the near-term problem. Future generations may face the same problem again. A method of destruction for asteroids and comets with collision orbits gives the only long-term solution.

The Russian Tungusta Event on June 30, 1908 has been the largest celestial event in recent history, and to the best scientific knowledge, the event was comet caused. Some evidence points to a fragment of comet Encke causing the nearly 800 square-mile, forest flattening disaster. A crater was never found, indicating the comet exploded a few miles above the forest. A 160 feet asteroid caused the Barringer crater in Arizona about 50,000 years ago. So, in recent times, the human race has had little to fear from the heavens, and until recent years, humans and countries have been quite lax or unconcerned about NEO or NEC that could wrought major life-changing destruction or possible extinction. For many years, we have neglected celestial watchfulness and countermeasures due to ignorance and lack of basic celestial understanding. However, the last few decades have allowed greater technology and knowledge of celestial happenings, and, to a lesser degree, we are thinking of NEO and ways to avoid catastrophes. There has been international cooperation, but the attitude and spirit of programs need emphasis. For this emphasis, bureaucratic bogs, procedural labyrinths, and counter country disputes need to be eliminated in lieu of a single purpose celestial disaster avoidance team composed of scientists, not politicians. The warning has been when, not if, and if the celestial certainty is not fully addressed with solutions, we need not concern ourselves about where we are going.

APARTHEID

How is it in modern times we have civilizations practicing discrimination of races, unless a class of people or a particularly evolved society seeks advantage or control for selfish motives? The motivation for such a sect in society must be high to subjugate races for control and self-aggrandizement. In South Africa the practice of apartheid has a long-standing history and the idea and function are contained within modern times where societies are, supposedly, intelligent and progressive. Modern times witnesses rampant discrimination, and in 1948 South Africa made it legal. Through the years the UN attempted various sanctions on South Africa, but initially the International Court of Justice ruled South West Africa was to be managed by South Africa, the apartheid proponent. Liberia and Ethiopia, later Namibia, comprised South West Africa. Full sanctions were never meaningfully applied, since the US, Great Britain, and other UN member countries sought strong economical trade with South Africa, and, primarily, for this reason, South West Africa would not become the independent country of Namibia until 1988. Economic prosperity has more control and power of longevity than the morality of abolishing apartheid.

Over the years, especially from the 60's to the 80's, many conferences, coalitions, committees, and anti-apartheid organizations were formed world—wide against the South African apartheid government and sympathizers. The US, Great Britain, and the Soviet Union were the most powerful countries to thwart apartheid. But in this mix, the Soviet Union wanted the economy of trade and the spread of communism in South Africa and surrounding countries economically dependent upon South Africa. The US and Great Britain did not enter into hard economic sanctions because each country also desired the economy of South African trade. Further, many British people lived in South Africa and were enjoying the white supremacy control, living conditions, and were sympathetic to Britain.

While some western countries had attitudes contrary to South African apartheid, they demonstrated only a psychological aversion to it with no positive action. The profitable economics of trade remain the stalwart motivation to resist anti-apartheid actions or sanctions. The consortium of the UN wanted to place strong economic and arms sanctions on South Africa, but in the 80's, the US and Great Britain wanted to follow a "constructive engagement" policy with the pro-apartheid South African government, believing it was a "bastion" of anti-Marxist forces. Certainly, South Africa did not want communism entrenched, but the bastion description was undoubtedly too strongly applied. Prime Minister Thatcher proclaimed one anti-apartheid organization within South Africa a terrorist organization. Still, the underlying reasons for western countries, including the US and Great Britain, for wallowing in rhetoric not solidly condemning apartheid were the economics of trade, monies, and international banking exchanges.

Slowly, international sentiment was gaining on anti-apartheid, but the white core supremacists of the South African government made considerable efforts toward dodging sanctions to include a development of nuclear weapons, supposedly with Israel's help. While South Africa did desist from a nuclear arms pursuit, the assistance from Israel has not been verified. Obviously and for some time, black opposition to the established white supremacists' South African government was growing stronger with its strength more than a match for the government to quell, but the supremacy attitude of the white leaders mandated against giving up as opposed to a complete dethroning of power and political reform. However, international pressures were growing, causing the South African government and President P.W. Botha to loosen, disregard, or repeal laws initiated by white supremacists. The effect of these actions and reforms were to seriously decrease the power and controls of the established white, previously powerful minority.

Although belatedly, other African nations began to actively contribute to the anti-apartheid movement. Reasons for delays were the strong dependencies of African countries to the economy of South Africa. These reasons still exist, but sentiment through the unfair discriminatory practices were strengthening. Nigeria's weight was beginning to be felt in the 70's and 80's, when it boycotted New Zealand's Commonwealth games due to New Zealand's nonchalant attitude toward the 1977 non-discriminatory Gleneagles Agreement. Again in 1986, Nigeria was a leader in a thirty-two nation boycott of the Commonwealth Games because Margaret Thatcher's rather neutral idea of sports with South Africa didn't support anti-apartheid. On this occasion and for a change, there was a significant economic input upon the games. The boycott had enough significance to receive international limelight.

From 1948 to 1994, apartheid flourished as authored by its designer, Doctor D.F. Molen, and in 1994 apartheid was abolished by South Africa, much to the happiness of many anti-apartheid countries. But, is apartheid still practiced? Evidence may be furnished by an adjacent township to Cape Town, Khayelitsha. Next to the riches of Cape Town are the poverty markings of overrunning, open sewers, rubbish heaps, unkempt streets, roaming cattle, and filth. Natural and man-made barriers separate the municipalities—apparently arranged. Structures strategically placed about natural barriers give the idea of international division, even after the official end of apartheid.

Although South Africa has been a center point for the practice of apartheid, Israel has been symptomatic of the human malpractice as well. The official Israeli government stance boasts no laws perpetuating apartheid, but observations of Arabs and Palestinians have shown otherwise.

Within Israeli history, there are some earmarks of apartheid. The Jewish government has been careful to identify Jewish nationals for preferential treatment or the advantage of material benefits. Possibly, such treatment was not intentional, but its repetitiveness indicates otherwise. The restrictions of keeping the Palestinians on their reserves, severing East Jerusalem from the West Bank, or the isolation of the Gaza Strip are governmental actions of apartness or apartheid. These actions separate non-Jews and Jews, a very discriminatory policy, and it has been noted by the outside world. For example, a United Nation's Special Rapporteur indicated apartheid in the Israeli occupation of Palestine, which is contrary to contemporary humanitarian laws. The administration of security intended to restrict Palestinians from assembly, freedom of movement, or the significant assembly of persons has been primary in Israeli non-Jewish personnel control. Suppression of dissidents to domination and control have also been the objectives of security.

Originally, the land of Israel was 96 percent occupied by Muslim and Christian peoples. But, for the Israeli government to maintain a preferentially Jewish state, apartheid measures must be applied, and they are—but not officially. Furthering the apartheid principal, the Israeli control of the West Bank and Gaza are outside the mores of humanity. The result of aggressive control are thousands of Palestinian citizens—children included—occupying Israeli prisons. Humanitarian needs are often denied. Examples are medical care and food. Although Palestine claims itself a separate state, Israel does not recognize its sovereignty.

The deeds of the Jewish government against Palestinians and non-Jews are of record and clear enough, but the irony of Israeli oppression reflects a lack of learning from history and a foresight for Israel. In the not too distant past, the Jews were suffering greatly at the hands of Aryan Nazis and for the same racial prejudices the Jews are now applying. Yet, the Jewish government aspires to an introverted view of itself and a sovereign, racial destiny. Where is the philosophical, humanitarian learning? Is a nation of people to suffer uncountable tragedies and yet not grasp or learn the significance of inflicting inhumane treatment upon others who do not follow their religion's guidelines or beliefs?

History has repeatedly shown nations or societies who suffer at the hands of others, yet, through a selfish introverted national sentiment, the recovering society feels compelled to inflict a similar, inhumane activity upon others. Inequalities have been established. At the least, Israel does not help non-Jews, and to a greater extent has applied apartheid activities or punishment. But, a hypocrisy exists in that the US is the largest foreign aid and arms contributor to Israel, while the US is theoretically against apartheid. For any nation, being true to deep rooted values can be difficult and, possibly, bring compromise to ideals.

Nuclear Weapons

Since the end of World War II, improvement upon and proliferation of atomic or hydrogen nuclear weapons have advanced. The Cold War advent from 1945 to 1992 with adversarial powers of the US and Russia were centered on deterrence of a nuclear war. During the Cold War, the numbers of nuclear weapons and operations of might and show were frequent. Militarily, US actions were correct and appropriate to counter or exceed Russian actions. Otherwise, the bullying actions of Russia would have meant a world takeover.

Fortunately, the Russian grasp for military might exceeded their financial and leadership capabilities, ending the Cold War nuclear mayhem. The Cuban Crisis was exemplary of near diplomatic failure for a nuclear war—fortunate for civilization. A point of observation needs emphasis: obviously, civilization has not advanced far enough for countries not to need the deterrence of nuclear weapons. While civilization produces nuclear weapons, the realization must be that these weapons are designed to obliterate a portion of civilization. All the auspices of the selected civilization are to be destroyed: literary works, sciences, medical facilities, and innumerable human advances. These things will be gone because the people, creators, of the society will be eradicated.

Until societies can coexist and have peaceful relations without leaders seeking control, power, and aggrandizement, nuclear weapons will continue to exist. As a forecast, leaders do and will seek control and power, and, unfortunately, these selfish traits require continuance of deterring nuclear weapons. The services of those who handle nuclear weapons will also be required. This is a deplorable but necessary scenario to temporarily foster world peace. The term temporary is applicable since mankind, as of the present, is not smart enough to pursue his long term existence in peace.

For an unforeseen time, nuclear deterrence will be embedded within civilization. Some politicians have come forth advocating nuclear weapons for certain countries. Adding fuel to the fire will not help the situation. For humanity's sake, what factor of extinction is satisfactory? During the Cold War, M.A.D., or mutually assured destruction, was the term statesman used to assure one another that no person would survive a nuclear exchange. Since it is impossible to forecast the tenure of deterrent nuclear weapons, a thousand years may not be too big of a span. The span is proportional to humankind's progress in the past, and society's retrograde in the last century supports a large span for societal progress.

Why is it that statesmen in eight countries have sequestered a multiplicity of nuclear weapons? Statesmen are supposed to be thinkers and leaders, and, apparently, they subscribe to the deterrence ideology. From a few hundred to several thousand weapons, some countries possess enough power for world annihilation. While power is a factor in the nuclear weapons deterrence scenario, delivery capability and accuracy are equally large requisites for effectiveness. The number of weapons in a country's arsenal for efficacy in deterrence may be a temporary necessity, but peaceful, productive negotiations might add longevity to civilization. If, unfortunately, negotiations and deterrence fail, then, and unknown offensive weapon count is in order, and at the same time, world—wide civilization will have failed. We should not be living that close to oblivion. Technologically, humankind has been intelligent enough to engineer weapons for our own extinction, but not resourceful or smart enough to draft a plan of peace, longevity, or world enhancing projects.

For consideration of current world dangers, a sovereign country should be scrutinizing North Korea and what its radical leader, Kim Jong Un, believes is best for the future of the nation. Kim's tribute to the country is embedded in his description of "magnificent" toward its nuclear weapon program. Kim's socialistic behavior was inherited from father, Kim Jong II and grandfather, Kim II Sung, indicating he had little chance for free world thinking, and his idea of diplomatic respect and world prominence is couched in nuclear provocations. What is involved in nuclear provocation? Will that be another attack on a sovereign country's vessel on the high seas, as has already occurred? According to Kim's nuclear philosophy, North Korea will be led out of poverty. This reasoning is fallacious, since radicalism does not incorporate logic.

Looking back on our wisdom and diplomacy employed during the Korean Conflict and early 50's, much of the trouble with North Korea could have been avoided had sterner measures been applied. Other more civilized and peaceable options were available for the Koreas. Obviously, past diplomatic errors cannot be changed, but the poignant past is recounted here to emphasize the errors couched in the concept of "politically correct" with which civilization is encumbered. North Korea has been, is, and will be a future problem for civilization. For all nations with nuclear power, long-range diplomacy should be the pattern for negotiations with longevity of civilization as the goal—not control, riches or power.

RELIGION AND SCIENCE

It would be difficult to place a date on humankind's arrival at conflicts between science and religion. Perhaps the conflict arose in the era when analytical thinking or scientific disciplines matured, since ideas of a god or a creator were in existence long before religious profits appeared. As scholars, scientists, astronomers, and mathematicians became more skilled, some revelations and discoveries became contrary to church dogma. Scientists did not intend confrontation, but some analytically derived data were unsupportive of church hierarchical teachings. In the early 1600's, Galileo Galilei proclaimed some observations with the telescope which were contrary to canon law of the Roman Catholic Church and the pope. From the meager beginnings of astronomy, Galileo intended no church conflict. He sought only to broaden the human perspective, which astronomically he did. But the church had already accepted Aristotle's and Ptolemy's proclamations as indisputable truths. With his celestial observations, Galileo was further amplifying a Nicholas Copernicus' heliocentrism theory which stated the earth and planets orbit the sun. With the Catholic Church's authoritarian stance of the world as the center of the universe, heliocentrism would question the church's authority on much established dogma and "truths."

Religion prophetically issues the values and guidelines of prophets and God for righteous, religious individuals to follow. These attributes of humans are not scientifically measurable; yet, these attributes are the essential differences between humans and other living forms. These attributes cannot be quantified but can be clarified as existent or nonexistent in humans. These values and guidelines are the essence of civilized humanity. They delineate between right and wrong or the goodness or badness of individuals within a civilization. In societies, mores and moral standards are developed with which individuals are judged. Scientists live within civilized societies, but there exists no credible way to scientifically measure the total morality or righteousness of individuals in society. Goodness may be exhibited on one day, while badness is evidenced on another. Can a resultant be calculated between good and bad? Humanity can proceed or regress in a sea of morality values. Yet, over the last few thousand years, humanity has progressed by many immeasurable values and factors, but the progress is difficult to measure or quantify. Tons of iron and barrels of oil produced can be measured, while the goodness of individuals producing the commodities cannot be measured.

When individuals become assets to the community through any application of biblical philosophy or religious teachings, a quantitative analysis of positive behavior is impossible to measure. Attributes that differentiate humans from animals: love, mercy, kindness, empathy, sorrow, curiosity, hate, and other esoteric human qualities, cannot be weighed, measured, or quantified to a specific unit. To advance beyond other species of animals, humans have developed brains to function with these attributes as well as thinking in terms of science, which requires analytical thought. Technologies with advanced civilizations followed. From the beginnings of Earth, to the ascension of Cro-Magnon, the probability of a human forming DNA combination seeks impossible—unless a creator intervened. From the discoveries revealed by science, eons have passed from early Earth to cognitive thought; whereby, time must be a tool of the Creator's. However, such a theory cannot be quantified or proven.

Science and discoveries of the Earth's assets and those of the universe, as we presently understand, are without limit. We simply know that we do not know—or do we? From empirical data and the tools of science and paleoanthropology, humanoids have probably been in their present form for 40,000 years or longer. From fossil records, Homo sapiens appeared about 200,000 years ago and, undoubtedly, coexisted with Neanderthals. A very low ratio of Neanderthal DNA still exists at less than 4 percent in modern man. In the years of two million years before present, there is evidence of weak cognitive processes, as Australopithecus robustus and Homo habalis probably used tools. If science and paleoanthropology are correct with the evidence discovered and analyzed, then, there existed evidence of humanoid progression—evolution. Is it necessarily evil that humans evolved from preceding humanoid forms? Or, did God advance humans in semi—sequestered steps through epochs such as the Pleistocene (2.6 million years ago) and the Holocene (11,700 thousand years ago). Answers are not readily available, but continual advancement of cognitive powers seem evident. Humans have been in their present physical form from long before the Holocene epoch, but given the span of time to the Miocene (23 million years ago) epoch, humanoids were probably nonexistent. Evidence produced by our own analytical thinking has given us reason humans have not always been in their present form. Is this discovery contrary to God's plan? No, it seems part of God's plan.

Religion, a philosophy developed by inspired individuals with wisdom derived from Prophets like Jesus or Mohammed, provides a metaphysical explanation to what cannot be explained by logic, reason, or science. Religious philosophy is deep and bears the attributes of cognitive powers, differentiating humans from other species. While scientists report the Big Bang occurred about 13.8 billion years ago and the earth formed about 4.65 billion years ago, they are working with the best tools and analyses available. There is always room for error. For those who have arrived at other theories of Creation and evolution through religious or Biblical understanding, there lies no fallibility or fault. Their cognitive reasoning produced a path to follow. There is no harm, as we must live in the present. However, about 60 percent of those following a religion believe in Earth's scientific date and an evolution process guided by God.

Science has a few facts to offer in a compatibility scenario with religion. When observing the periodic table of elements, how is it that a specific element has a specified number of protons, neutrons, electrons? Changing the number of protons will change the element. For example, carbon has a proton count of six and an atomic weight of a little over twelve. Therefore, carbon has an atomic number of six, an electron count of six, and six neutrons. On earth, carbon is the staff of life, but by deleting or adding one proton to carbon, we have a different element. Delete one proton and the element is boron, and by adding one proton, the element becomes nitrogen. The periodic chart, generally, follows elemental change by proton change. Without guidance, the probability of element organization by protons approaches astronomical.

By looking at scientific discoveries, those following religious beliefs can see science correlates data gathered by observation. We are slowly learning about the earth and the universe by discovery. Earth's gravity has always been present, and Sir Isaac Newton was the first to describe and quantify its behavior. Later, Albert Einstein modified the description. To synopsize, humans discovered the laws of geometry, trigonometry, algebra, and calculus, while similar discoveries are applicable to chemistry, physics, thermodynamics, and many more disciplines. Inventions occur regularly through the utilizations of these scientific disciplines. Most of the laws or rules of science were discovered; they always existed while awaiting discovery.

As humans apply science to learn their place in the solar system, galaxy, and universe, probability would be against the Earth's ideal, life-giving celestial position, unless guidance intervened with time. To foster life, the earth is at an ideal size. Without our notice, the earth is large enough to hold an atmosphere and water, the essentials for life. If the earth was a great deal larger, gravity for locomotion could be a problem. In terms of solar system position, the earth is in the ideal location. A position closer to the sun would produce a hot uninhabitable planet devoid of an atmosphere and water, and a position farther away from the sun than Mars would produce an icy, frozen, uninhabitable world.

From about 4.45 billion years ago an earth impactor, Theia, collided with Earth to provide ejecta materials to coalesce into our present day moon. This scientific theory is not perfect, nor does it explain all questions, but it does provide an idea for the moon's formation. From what evidence we have, the moon and its materials are identical to that of the Earth's. Regardless of the moon's formation, it is most beneficial and life-giving to earth. More than likely, the impactor provided the moon, Earth's rotation, and Earth's tilt. With these attributes, humans receive regularity, tides, seasons, and a predictable day. After the collision, Earth's days were much shorter, five or six hours, but with the passage of over4 billion years, humans have a beautiful twenty-four-hour day. Did the Creator use a tool of time to produce a utopia for humans to evolve?

Science does not have a clear explanation for the evolution of oxygen, O_2, which is a life-giving and sustaining element. From the geologic records, oxygen was only about .01 percent of the Earth's atmosphere at 2.7 billion years ago and, maybe, .1 percent at 2.45 billion years ago, but it became and remained part of the Earth's atmosphere. Sometime in the early Triassic Period cyanobacteria took hold and produced enough oxygen to enable evolution of animals. The oxygen building process took until the late Jurassic Period for significant quantities. Through the continuing cyanobacteria process and other earth shaping features such as tectonic plate drift, oceans forming, volcanoes, and trees developing, the current oxygen level arrived at twenty-one percent. So, it took about 2.7 billion years to get our current, livable oxygen level. Was this long, step-by-step process planned, or did it just happen? Science's predilection tends toward an advancing, life enhancing plan.

Ozone was discovered in 1839, but it's chemical formula of O_3 was not known until the Civil War. The scientist proclaiming the formula was Christian Schonbein. It is not known when substantial concentrated levels of ozone were present in the Earth's stratosphere, but the period must have been after plentiful oxygen was present in the late Paleozoic or early Mesozoic eras. The substantial rise of plants and animals needed greater oxygen levels, but each life form needed protection from the sun's harmful ultraviolet rays, which could damage DNA. The reaction of ultraviolet rays with the two molecule oxygen, O_2, would produce ozone, O_3, a stratospheric protective layer to earth life forms.

But ozone at the ground level is harmful to animals and plants. At low atmospheric levels, ozone is a pollutant. It can interfere with lungs and bronchial tube efficiency as well as plant photosynthesis. Life needs the benefit of ozone's presence but not in the living forms. Science has given humans the knowledge of ozone's presence, the benefits of ozone, and the detrimental effects of contact with it. But without the benefits of ozone, could higher life forms such as humans have evolved? Again, answers are not readily available, but animals in the Mesozoic era probably would not have evolved to their late Cretaceous Period state it ozone was not present.

With what paleoanthropological knowledge we have, it does not appear the creation and evolutionary processes superseded the laws of nature or provided something for nothing. In the processes of time, a Creator must have provided the catalyst of ozone. Science has not contradicted the existence of a Creator but has revealed the time woven intricacies of a guiding hand to our universe and world.

ABOUT THE AUTHOR

Robert E. Barr was a pilot for over fifty-one years and flew in both the military and civil environments. Military flying included three tours of combat flying in Vietnam that earned the Distinguished Flying Cross, DFC, and fourteen Air Medals.

Bob's civil flying included flight instruction, but a great deal of his civilian flying was that of a corporate pilot for an oil service company with Dick Cheney as one of the CEO's. Oil search and service endeavors encompass the world to be successful, and the experience gained from world-wide flying proved invaluable for perspective. Flying over and working on foreign soil broadens international relations understanding and negotiations skills.

Bob earned a BS degree from Wichita State University, an MBA degree from Oklahoma City University, and an MA degree from the University of Northern Colorado. His flying credentials include an airline transport pilot certificate, ATP, a commercial pilot certificate, instructor ratings, type ratings, and an air traffic control certificate.

Bob lives in Boise, Idaho